T0320408

Generative AI and Cyberbullying

Ever since the COVID-19 pandemic occurred in 2020, the world has transformed itself greatly. For example, not only is the near-99% remote workforce now a reality, but businesses today are taking incident response and disaster recovery much more seriously these days as well. But another area that has blossomed in the last couple of years has been that of Generative AI. It is actually a subfield of artificial intelligence, which has been around since the 1950s.

But Gen AI (as it is also called) has been fueled by the technology of ChatGPT, which has been developed and created by OpenAI. Given the GPT4 algorithms Gen AI is powered by, an end user can merely type in, or even speak into the platform a query, and an output that is specific to that query will be automatically generated. The answer (or "output") can be given as a text, video, image, or even an audio file.

The scalability and diversity of Gen AI have allowed it to be used in a myriad of industries and applications. But although it has been primarily designed to serve the greater good, it can also be used for very nefarious purposes, such as online harassment and Cyberbullying.

In this particular book, we actually take the good side of Gen AI and provide an overview as to how it can be used to help combat Cyberbullying. This book is broken down into the following topics:

- What Cyberbullying is all about
- How Gen AI can be used to combat Cyberbullying
- An overview into Gen AI
- Advanced topics into Gen AI
- Conclusions

Ravindra Das is a technical writer in the Cybersecurity realm. Additionally, he also does Cybersecurity consulting through his private practice, M L Tech, Inc. He also holds the Certified in Cybersecurity credential from the ISC(2).

Generative AI and Cyberbullying

Ravindra Das

CRC Press
Taylor & Francis Group
Boca Raton London New York

CRC Press is an imprint of the
Taylor & Francis Group, an **informa** business

Designed cover image: © Shutterstock

First edition published 2025
by CRC Press
2385 NW Executive Center Drive, Suite 320, Boca Raton FL 33431

and by CRC Press
4 Park Square, Milton Park, Abingdon, Oxon, OX14 4RN

CRC Press is an imprint of Taylor & Francis Group, LLC

© 2025 Ravindra Das

ISBN: 978-1-032-66600-6 (hbk)
ISBN: 978-1-032-66601-3 (pbk)
ISBN: 978-1-032-66602-0 (ebk)

DOI: 10.1201/9781032666020

Typeset in Caslon
by Apex CoVantage, LLC

This book is dedicated to my Lord and Savior, Jesus Christ, the Grand Designer of the Universe, and to my parents, Dr. Gopal Das and Mrs. Kunda Das.

To my loving cats, Fifi and Bubu

This book is also dedicated to

Richard and Gwynda Bowman

Jaya Chandra

Tim Auckley

Patricia Bornhofen

Ashish Das

Contents

Acknowledgments

I would like to thank Ms. Gabrielle Williams, my editor, who made this book into a reality. I would also like to thank Gilles Pilon for his contribution to this book.

Gilles Pilon is a professional engineer, Python developer, and data scientist. He started his career as a process engineer in aluminum rolling mills and transitioned to data science in the food industry and then to artificial intelligence and machine learning.

1

INTRODUCTION
My Experiences as A Bullying Victim

As of this writing, this will be my seventeenth book. All of my books in the past have ranged from biometrics all the way to Cybersecurity. They all have been very technical in nature, talking about the mathematical equations used in encryption to the algorithms that are used in artificial intelligence, machine learning, computer vision, neural networks, natural language processing, the digital person, Generative AI, etc. A common denominator in those books was that I never shared my personal "feelings" into them. But this book is going to be a little different.

Rather than starting to talk all about the technicalities of what this book is about, I am going to discuss some of my life's history and what brought me to write the topic for this particular book. I was born in West Lafayette, IN, to immigrants from India. My father was a professor of biological sciences at Purdue University, and his primary area of research was spinal cord regeneration. He was a leading pioneer in the field, even having won the ever prestigious Jacob Javits Award. But he died in December 1991 of a massive heart attack.

Before all of these happened, he was currently working on his third book and was apparently on the verge of some sort of breakthrough in spinal cord regeneration research. Apparently, he discovered a new way in which to use amniotic fluid to help the transplanted cells live longer in the spinal cord. The results he had obtained were preliminary but had great promise for potential going forward into the future. My mother also worked at Purdue University, along with my father, helping him with both his research and writing/publishing scientific journal articles. But she too died of cardiac disease, not being able to survive a second round of heart bypass surgery.

DOI: 10.1201/9781032666020-1

My parents brought me up in the way that most parents would their own kids. They were extremely loving, supportive, and patient in everything that I had tried to accomplish growing up. But one thing about their upbringing to me is that they sheltered me to extremes. A lot of things that I should have experienced I never did, such as dating in high school and even going to the high school prom.

But despite all of this, there was one common denominator about me that totally stood out from most kids being raised. I was mercilessly bullied by everybody. I remember it all started I think back even as far as fifth grade. I remember I was late in starting the term because I had taken a summer trip to India with my sister. When I entered the classroom, all of my classmates just looked at me, and one even dared to make the comment: "Everything was going so great until he showed up".

I don't remember too much about the bullying back then, but I do remember this one particular boy, who shall remain nameless, always picked on me and beat me up all the time. I had him as a classmate from grades 5 to 6, and his incessant torment never backed off. I thought that there were times we would be friends, but that was just a trick to lure me in for more abuse. It got so bad one time I remember that my mother actually had a talk with him about all of this bullying I was getting, and even the elementary school principal talked to him about it. But nothing ever changed; in fact, it made just the other classmates pick on me that much more.

Oh well, at this point, elementary school was now over, and now I had entered another era of pure hell, which was known as junior high school. All I remember that seventh grade was the hardest year I had experienced in my schooling so far, and a lot of this was brought on by my science teacher. She came from Germany (actually more formally known as "West Germany") during this time; so of course, she had extremely high expectations out of all of her students. Unfortunately, I could never make the grade, and she too picked on me as well. Of course, as an educator, she had certain lines that she could not cross with regard to this, but in her own way, she knew I had potential, and just wanted me to tap into it. The way that she did it was completely way off, but it was not until closer when school was out did I actually realize what she was trying to do.

So with that mind as motivation, I studied as hard as I could, and my grades in her class really started to pick up. As the crowning achievement, on the very last day of school before we were let out for summer vacation, she actually gave me a hug and said: "I knew you could do it. I am so proud of you". So with that new spark in me, eighth grade started a short time later, and I had some of the best teachers I ever had so far in my schooling (with the exception of my math teacher).

I don't really remember being bullied that much in the eighth grade, but as the next four years of high school started, these were among the worst days of my life. In fact, I remember my father even saying that "high school are amongst the worst years of anybody's life". And, he was so right about that. During the next four years of this torturous life, I was picked on almost every day and constantly bullied. I don't remember each and every single day, but the one thing that I do remember was that there were these three boys who bullied from morning when school started until after we let out in the afternoon. They were big boys, all of them played on the football team.

So of course, they needed someone to bully to show off their prowess to everybody else, so they naturally picked me as the target. I think a large part of the reason for all of this horrible treatment was that I was also over weight. I would say that I was not morbidly obese, but I could have lost some serious weight. So, now fast-forward to my junior year in high school, my parents in an effort to help lose some weight, encouraged me to join the boys swim team. The swimming coach was great and was probably one of the nicest and most caring individuals that I have ever come across in my entire life, even up to this point as I write this book.

But unfortunately, the boys on the team were not so nice. I was picked in a weird way, in the sense that the four seniors who led the team victory that year did like me, and they referred me to as their "pet". Deep down, I knew that this was wrong and completely degrading. But, I accepted this, in an effort to forge a new type of friendship that I never had before. But the boys in my class and younger were horrible in their bullying to me. In fact, the same boy who bullied me from grades 4 to 6 in elementary school also went to the same high school, and at that point, the picking on got even worse. I was spit upon and even urinated upon, but I never told my parents about this.

My grades took such a nosedive that I remember in my junior year, I had even received a few "F"s in subjects on my final transcript. In fact, it got so bad, my parents lost total hope of me going to Purdue, but they were now fearful I would not even graduate from high school at all. The bullying continued, but I had also decided my grades had to get better. So it was not until the fourth grading period of that year that things actually started to improve. I do remember at this point in time also there was yet another boy who was far worse in his bullying, and even tried to stab me in the kidneys with his pocket knife.

So, now we come into my last final time in high school, which was senior me. Now with my grades starting to show some improvement, my parents were not so much worried about me graduating from high school. Now, it was if I would be able to get admitted to Purdue, or be forced to go to a junior college in the interim. So in order to accomplish this goal, there was no more being on the swim team, and my guidance counselor now took even further efforts to make sure I would do well by having me take some easier courses. It all started to pay off. That first semester, I had landed a solid "B" average, and even got some "As" in there as well. I was still being bullied though, but not nearly as much as in my first three years of high school. My grades continued to do well, and in the last semester before graduation, I landed yet another solid "B" average. In my junior year, I was ranked toward the near bottom in my class, but with this huge uptick in my grades, my class rank greatly picked up, and in fact, I was now ahead of at least 35 other students.

So in the end, I finally graduated, and a few weeks later, the best news came: I now actually got accepted to Purdue. When compared to other college students, it took me longer to graduate; in fact, it was almost six years total. The good news here is that a lot of bullying that happened in high school did not happen in college. But the reason why I took so long was that going to college was such a different experience for me than merely going to high school. In fact, I switched majors something like five times before I settled upon something that I really liked, which was agricultural economics.

My father died about a year before I graduated, so it was just me and my mother now during this time period. I really wanted to get a job, but my mother really wanted me to go onto graduate school to get my Master of Science in agricultural economics. My dream was to go to Purdue again, but unfortunately, I did not accepted. In the end, I got accepted to Southern Illinois University, Carbondale (also known as "SIUC"). I was actually very excited about going, and to make my mother happy, and above my all, my father.

But now something else happened to me yet again, for which I was not ready. Yes, another round of bullying was about to start. But this time, it would be an entirely different kind of one, based on racial discrimination. You see, coming from Indiana and being completely sheltered by my parents, I knew nothing of what the term "Red Neck" meant. And, I learned this the hard way when I went to SIUC to start my graduate school days. Most of the "White and Southern" students were given a nice and elaborate office area, whereas the foreign students were given something much smaller.

Luckily, I was able to get a desk space where the "White and Southern" students' office was present. During my first semester there, I did my best to make friends with everyone, and in the end, most of the "White and Southern" students started to finally accept me. But there was one student who still kept bullying me, but he finally stopped around sometime in the second semester. Also keep in mind that this was my first time living in a dormitory, so I got picked on there as well, but that too eventually faded away as I got acclimated to that kind of environment.

I took a job as retail manager of a grocery store/pharmacy store. But unfortunately, that did not work out too well, and I was fired. So now during this time, I had to face yet another challenge as to what I wanted to do with my life. In the end, I decided to get my MBA from Bowling Green State University (also known as "BGSU"). But yet again, the bullying picked up. This time, it was not with the "White" group, it was, hard to my belief even today, with the "Indian" group in my class. I can't remember all of the details per se, but I remember I had the hardest time getting to be a part of a group for the group projects, and when I finally was picked, I was taunted mercilessly about my weight, as I still had my poundage back from high school, and even more.

But it did not stop there. During my MBA program, I had also taken a few trips to India, as my relatives had a business there, and they wanted me to help them expand into the US market, after I had graduated. Now, while these specific relatives did not taunt and tease me, the others did, me being overweight, and it was among the worst form of bullying that I have ever experienced in my life. I kept thinking to myself while I was in India: "Why am I even here? What I have done to ever deserve this?" As a result, because of all of this, I have never returned to India and don't have plans to in the future.

So now, you the reader are probably thinking this: "Why is the author bringing all of this up? Is this book an autobiography?" Well, the primary reason why I have brought all of this up to you is that this book is not about myself, rather it is *about bullying, and even more important, Cyberbullying.* As I had mentioned at the beginning of this chapter, I could have just started with all of the technicalities of Cyberbullying. But instead, I wanted to bring up my own experiences on this, to show how real and devastating it can all be.

The reality of this is also there is no formal legislation or punitive action one can take in order for them to stop from becoming a victim of it. It is not just kids who are being bullied, but it can even happen to adults at all ages, and even in all sorts of differing environments. But what has made this even worse is the dawn of the digital age. Today, we live in a world that is ultra-connected, brought on by many things, such as the internet and the internet of things (also known as the "IoT"; this is where all of the objects that we interact with on a daily basis are connected together, both virtually and physically).

Because of this, a person who is located literally thousands of miles away can harass, taunt, tease, humiliate, ridicule, and even threaten their victims, primarily through the use of the various social media platforms (such as TikTok, Instagram, Pinterest, LinkedIn, X (formerly "Twitter"), and Facebook). Yet, there is very little that can be done about this horrible situation.

Another way of looking at this scenario, bullying is one thing when you actually confront the perpetrator(s), but it is an entirely different thing when you cannot even see what they look, and they can even be in an entirely different country for that matter. Therefore, the goal of this book is to address what Cyberbullying is actually all about, the impacts that it has on the different victims, and the different manners in which Cyberbullying actually occurs in.

But, this book also takes an interesting turn. We had just mentioned that technology is the primary culprit for Cyberbullying, *but we now examine how it can used to actually combat Cyberbullying.* In this regard, we will be taking a closer look as to how artificial intelligence and its subfields can be used to help accomplish this particular task.

1.1 What Is Cyberbullying?

The concept of "bullying" probably has existed since the dawn of mankind. But as one can see, it has evolved over time into much more sophisticated levels than one could have ever imagined. But, it is important to make the distinction between "Traditional Bullying" and now the more common "Cyberbullying". A technical definition of "Traditional Bullying" is as follows:

> Traditional bullying typically refers to aggressive behavior that takes place in in-person settings, such as schools, neighborhoods, or workplaces. The hurtful behavior is direct and the bully interacts face-to-face with the victim.

> **(SOURCE: https://www.dispartilaw.com/what-is-cyber bullying-definition/#:~:text=Traditional%20bullying%20typically%20 refers%20to,physical%20aggression)**

Some of the best examples of "Traditional Bullying" include kids pushing and shoving each other on a playground, or perhaps even in the classroom, if the teacher actually allows it. But it is important to note that "Traditional Bullying" does not actually have to come in the form of physical contact. For example, many experts in this area consider the following to be forms of "Traditional Bullying":

• Verbal abuse
• Social exclusion
• Spreading rumors

But a key differentiator here is that while you do not have to physically altercate an individual, the person must be in close enough proximity to you in order to deliver the following examples. For example, in my story that I provided at the beginning of this chapter, the bullying that took place in grades 4–6 did involve some sort of physical abuse. This lessened in high school to a degree, and in college and graduate school, there was no type of physical contact, but the above examples still were very much present. Therefore, even under these circumstances, I would still be considered to have been bullied in the traditional sense.

Now, as I had also mentioned earlier in this chapter, given the advent in the explosion of wireless devices and virtualization, most forms of bullying take place now online, but it is also important to keep in mind that many variants of "Traditional Bullying" still exist on a global basis and will continue for a long time to come.

So, this brings us up to our next question: "What Is Cyberbullying?" This term can be technically defined as follows:

> Cyberbullying is the use of technology to harass, threaten, embarrass, or target another person. Online threats and mean, aggressive, or rude texts, tweets, posts, or messages all count. So does posting personal information, pictures, or videos designed to hurt or embarrass someone else.

(SOURCE: https://kidshealth.org/en/teens/cyberbullying.html#:~:text=Cyberbullying%20is%20the%20use%20of,hurt%20or%20embarrass%20someone%20else)

Thus, the non-contact forms of Traditional Bullying are in pure existence with that of Cyberbullying, but instead of being in close proximity with the victim, there is no physical interaction between the perpetrator and the victim. Also, rather than having an actual, physical surrounding be the venue for Cyberbullying, it all takes place over the internet, and into the wireless devices of the victim. As it was also mentioned at the beginning of this chapter, Cyberbullying most often takes its root and delivery via the social media platforms that are present today.

An example of "Traditional Bullying" can be seen in Figure 1.1:

Figure 1.1 An example of "traditional bullying" on physical premises.

SOURCE: https://www.shutterstock.com/image-photo/teenager-bully-menacing-boy-while-friends-2158259463.

An example of "Cyberbullying" can be seen in Figure 1.2:

Figure 1.2 An example of Cyberbullying on a wireless device.

SOURCE: https://www.shutterstock.com/image-photo/woman-sit-on-sofa-holding-smartphone-1761076589.

There are also other, subtle differences between "Traditional Bullying" and Cyberbullying, and this is reviewed in more detail in the next section.

1.2 The Differences between Traditional Bullying and Cyberbullying

It should be noted that the two common characteristics of "Traditional Bullying" and Cyberbullying are that of aggressive behavior and actions (which could also include physical violence to any kind or degree), and the willful intention to cause harm or any sort of distress onto the victim. But, here are some of the differences:

1. *The Setting*:
   With "Traditional Bullying", as mentioned, the venue is usually some sort of physical setting. Once again, it can include the playground for kids, the streets and buildings for city-based gangs, and even the physical location of a business if "Traditional Bullying" takes in place in a work-like environment.

   But with Cyberbullying, since it is all done through the internet, the world is literally the venue, to wherever the location might be that the victim has an internet-based connection.

2. *The Effects*:
   The effects of the abuse by bullying in both cases that are felt by the victim will course depend upon them and their unique circumstances (this will be later reviewed in this book). But generally speaking, in terms of "Traditional Bullying", the effects are usually felt for a much longer time period. The primary reason for this is both the physical contact that they may have been involved and the close proximity that exists between the perpetrator and the victim.

   But with Cyberbullying, although the effects can be just as bad, they are usually not as long lasting. The primary reason for this is that the victim can simply "block" the victim from the social media platform that they are using, or they report the perpetrator directly by filling out a form online. But the caveat here is that if the perpetrator is actually very obsessed with their victim, they can simply create another profile on that particular social media platform and use a different alias as their name and username. In the end, the same victim will not realize that they are being followed yet again by the same perpetrator, until the telltale signs start to emerge yet again.

3. *The Victims*:

One of the other common denominators between both "Traditional Bullying" and Cyberbullying is that there are victims whom will be impacted in both the short term and the long term. But how they deal with that will be extremely different from person to person. In terms of "Traditional Bullying" the actual total number of victims that are involved is actually much less. Unless there is a large gang that is involved in the particular event, there is usually only a few victims that are impacted in these cases.

But on the contrary, the victims of Cyberbullying will of course be much greater, because as just stated, using the old proverb, "the world is the oyster" for the perpetrator. But what is also important to note here that while there can be hundreds, thousands, or even millions of victims that can be impacted, the speed of delivery is obviously far quicker than that for "Traditional Bullying". For example, a "Deepfake" video can be created about a person, and literally go "viral" almost instantly in just a matter of seconds. It should be noted here that "Deepfakes" are actual fake images or videos of real human beings, in order to target them in a Cyberbullying attack. These are created by using artificial intelligence.

4. *The Sense of Anonymity*:

This is deemed to be one of the most important differences that exist between "Traditional Bullying" and the Cyberbullying. With the former, the victim can actually see the faces of their perpetrator(s), unless of course, they are wearing a mask. Also, with the advent in the use of CCTV cameras in conjunction with facial recognition, the faces and even the bodies of the perpetrator(s) can be recorded, and even analyzed on a very sophisticated level by law enforcement in order to bring them to justice.

But with Cyberbullying, there exists an extremely high level of "anonymity". This simply means that the victim will never truly know who their perpetrator really is, since they are being assaulted many thousands of miles away, in an entirely different country. Also, the perpetrator can easily create an alias of their name, and even upload a phony picture of themselves, once again, using Deepfakes. The only true way that they even be possibly recognized is if law enforcement can actually track them down, using primarily the principles and methods of digital forensics. But in actuality, the chances of this really happening are quite remote, as the degree of the severity of the Cyberbullying has to be extremely grave, and the victim must be impacted consistently for a very long period of time.

More differences between "Traditional Bullying" and Cyberbullying can be accessed at the following link:

http://cyberresources.solutions/Cyberbullying/Cyberbullying_
Traditional_Bullying.pdf

1.3 The Stages of Cyberbullying

Now that we have discussed in some detail what "Traditional Bullying" is and how it varies from Cyberbullying, the main emphasis for the rest of this book will be primarily on that of Cyberbullying. In this kind and type of attack, the stages in which it typically occurs are as follows:

1. *The Victim*:
   Although a Cyberbullying attack can involve anybody ranging from the youngest of children to the oldest of senior citizens, the perpetrator needs to have a victim. Now while they pick this person at total randomness, the Cyberattacker of today is very careful and deliberate as to how they pick their victims. The trend now for them is to take a much longer

time period to pick out they will specifically target. But when it comes to Cyberbullying, the perpetrator will often study everything that they can about their victim, using whatever publicly that they can. This can be done ranging from using what is known as using the tools from "Open Source Intelligence" (also known as "OSINT") to even conducting anonymous, detailed background checks. One of the primary reasons why the Cyberattacker takes so long to target their victim is that they also want to discover whatever weaknesses that they can about them. After all, this is one of the "best ways" possible to make a lasting and harmful impact upon the victim.

2. *The Action*:

   In this phase, the perpetrator is now ready to launch their Cyberbullying attack. Before this, they will also have chosen what they will deliver, and the specific venue in which it will be delivered in (most likely a social media platform, or even a combination of them – if the victim has multiple social media profiles). The intention now is to create as much damage as possible, perhaps even going all the way to an extortion-like attack in which the victim is forced to pay money.

3. *The Repetition*:

   In a traditional Cyberbullying attack, it does not just happen once. It usually happens over several iterations, in order to create the maximum of negativity that is possible. But typically, they are done over gaps of time, in order to catch the victim off guard. Also, it is quite conceivable that the perpetrator can launch multiple attacks across the various social media platforms that the victim is on.

4. *The Imbalance*:

   Once the Cyberbullying attack has been launched and has been done over numerous iterations over a long period of time, the perpetrator will only feel satisfied once they have complete control over their victim, even if it is from thousands of miles away. Once this happens, the phenomenon known as the "Imbalance of Power" has now been finally reached.

These specific phases are illustrated in Figure 1.3:

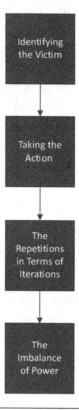

Figure 1.3   The major steps in the launch of a Cyberbullying attack.

1.4 The Differences between Cyberbullying, Cyberstalking, and Cyberharassment

Now that we have provided more detail and information into Cyberbullying, it is important to note that there are other variants of Cyberbullying as well. What the exact classification is of them depends largely upon the extent of the act, and how long it lasts. According to researchers, the following are major different variants of Cyberbullying:

1. *Cyberstalking*:
   This is where unwanted messages are sent repeatedly to the victim. But in this specific case, the perpetrator (or even

the Cyberattacker as we have mentioned to in the past) will even go so far as to threaten the victim with their life. Or, an extreme case of this is when, if the perpetrator is close by, they will even come into some close proximity of their victim, and he or she will not even know anything about it what is actually happening. This is usually a prelude of something more dangerous happening at a subsequent point in time.

2. *Cyberbullying*:
   This is what we have examined and defined thus far. The signs of Cyberbullying include sending out what are known as "Memes". This can be technically defined as follows:

   A meme (pronounced meem) is a piece of content – usually an image, video or text – that is typically humorous or sarcastic in nature. Popular memes spread rapidly online, with different variations circulating, trending and being recreated.

   > (SOURCE: https://www.adobe.com/express/
   > learn/blog/what-is-a-meme)

   But the difference here is that when it is used in the context of Cyberbullying, the "Meme" is typically designed and created in order to cause harm onto their victim. The act of actually doing this is called "Memeing". An example of a "Meme" can be seen in Figure 1.4.

3. *Cyberharassment*:
   This can also be considered as a form of Cyberbullying, but the key difference here is that the perpetrator will most likely just submit a few nasty remarks or comments about their victim onto a social media platform, and simply let it go at that. Also, the duration of this kind of Cyberbullying is much shorter, in terms of a time frame.

So in terms of all of these variants, Cyberstalking is considered to be the worst form, Cyberbullying is "mild" in terms of damage, and Cyberharassment is deemed to have the least amount of impact onto the victim.

Table 1.1 further details some of these differences.

Figure 1.4   An example of a "Meme".

SOURCE: https://www.shutterstock.com/image-photo/funny-friends-watching-media-on-phone-2365987919.

Table 1.1   The Various Methods of Online Harassment

CYBERSTALKING	CYBERBULLYING	CYBERHARASSMENT
Repeated, unwanted messages sent over a long period of time	The incidents are fewer, but repeated, unwanted messages are sent, and can be for a long period of time	Unwanted messages are sent only periodically and usually stopped in a short period of time
The victim is fear for their life	A hostile environment is created, but the victim is not necessarily in fear of their life	The victim is not necessarily threatened, but they are belittled and ridiculed
There are Federal Laws against this which can be enforced in a Court of Law	There are no Federal Laws, but many states have laws, which can also be enforced in a Court of Law	There are no Federal Laws or State Laws specifically, and Cyberharassment is only illegal if it starts to fall into the categories of Cyberbullying and/ or Cyberstalking

1.5 The Evolution of Cyberbullying

Many researchers have their own, differing views as to how Cyberbullying actually got started. As we have mentioned earlier in this chapter, Cyberbullying is actually an evolution of Bullying, which has probably existed for the past thousands of years. But, the consensus seems to be that Cyberbullying actually started to proliferate in the early 1990s. This is when technology really started to take off, driven primarily by the internet bubble, also known as the ".com Bubble".

It was not only computers that advanced quickly, but even the growth and adoption of wireless devices as well. As a result, many chatting and messaging tools became available, and because of this the Cyberattacker now saw their chance of harassing their victims from a secret location many thousands of miles away. But when the internet bubble started to burst by 2000, a new evolution started to take place. This gave birth to the "mobile apps".

With this new advent, the Cyberattacker now had an even further reach to target and harass their victims. They did not have to wait to see when their victim would come online to whatever social media platforms were present back then, rather they could send unwanted and threatening messages to their victim straight to their mobile apps. In fact, mobile apps are now such a way of life for end users; this is probably now the best way for a Cyberattacker to cause the pain and misery that they wish to inflict upon.

The next trend in Cyberbullying started in 2010. This is when the social media platforms now started to create their own mobile apps, so that the end user would not have to log in directly from the web interface. This, of course, was yet another major catalyst for the Cyberattacker to further amplify their attacks onto their victims. This trend has even continued to this day and has gotten far worse, given the advancements that are being made in mobile apps, and the interconnectivity among people of their wireless devices.

In fact, it was during this time period that one of the first pieces of Federal Legislation against Cyberbullying was passed in 2010. It became known as the "Tyler Clementi Higher Education Anti-Harassment Act of 2021". This new Law mandated that any colleges or universities that were receiving any type or kind of Federal Aid now had to put into place anti-Cyberharassing ordinances in order to further protect their students. These major highlights of the evolution of Cyberbullying are illustrated in Figure 1.5.

Figure 1.5   The evolutionary timeframe of Cyberbullying.

1.6 The Different Types of Cyberbullying

From within the category of Cyberbullying, there are many different variants of it, depending upon which the perpetrator (or Cyberattacker) wishes to employ. In this section, we examine some of the major ones that people, no matter what the age is, typically encounter, if they do indeed become a victim of Cyberbullying. They are as follows:

1. *The Social Exclusion:*
   This is deemed to be one of the most common forms of Cyberbullying. To give an example, as it relates to "Traditional

Bullying", this is where an individual feels left out in a group setting. For instance, it could be a young child left alone in a certain area of the playground, while their classmates play in a different area of it, or an adult being left from a group of peers, no matter what the environment or the occasion might be. This is what I had experienced when I was in graduate school and was further described at the beginning of this chapter.

2. *The Outing*:

   This is also technically referred as "Doxing". This is where the perpetrator typically finds out some personal or private information about their victim and shares it with others without the prior consent of their intended victim. The primary objective here is to spread this private and confidential information as a rumor in order to cause and inflict harm to the intended victim. This is done on a high basis on the social media platforms, mostly on Facebook and X.

3. *The Trickery*:

   This technique can be considered to be the same as the last one. But in this particular instance, the perpetrator will make themselves known to their victim, and actually even befriend them over a period of time. This is all in an effort to con the victim and use that weakness to launch threats against them, or even extort them for money.

4. *The Fraping*:

   This is actually an acronym that stands for the combination of "Facebook" and "Raping". This variant first made its legacy on Facebook and then it eventually spread over to the other major social media platforms. In this instance, the perpetrator takes over the entire account of the victim and posts malicious things about them. In this case, the victim has no longer control of their account and as a result will have to create a new one. It should be noted here that today, the social media platforms take this very seriously, as it can be considered as an act of a Cyberattack.

5. *The Masquerading*:

   This act of Cyberbullying is considered to be similar to "Fraping". But in this case, the perpetrator does not take direct control of the victim's social media platform account; instead, they recreate a duplicate of it, but with fake contact information. Also, the perpetrator tries to replicate the content, with some changes, in order to con the actual social media platform of the victim's friends.

6. *The Dissing*:

   In this kind of attack, the perpetrator tries to affect the relationships that the victim has with their social media platform accounts. In order to make this kind of threat successful, "Misinformation", Rumors, and other types and kinds of Falsehoods are spread around about the victim. This is also known as "Denigration".

7. *The Trolling*:

   In this particular case, the perpetrator will often post negative statements and comments about their victim, in an effort to get others to also go against him or her. What makes this different from the other Cyberbullying variants is that typically, the Cyberattacker (also the perpetrator) will not even know the victim, and the others who go against the victim as well will not even know of them. This kind of attack typically takes place on where "Threads" are used as the main forum, such as X (formerly "Twitter").

8. *The Roasting*:

   This is also referred to technically as "Flaming". This is where the perpetrator does serious damage to their victim by using profane language, in a very abusive form. Once again, this is often done on the social media platforms, or other major, online forums.

9. *The Impersonation*:

   In this case, the perpetrator directly emulates the victim, by following them closely as to what they have posted on their social media platform accounts. In fact, the Deepfakes, in order to create a replica picture of the victim, is often used here as well.

10. *The Catfishing*:

    This is where the perpetrator preys upon the emotions and feelings of the victim. A typical platform used here are the online dating sites. Typically, the victim feels attached to their perpetrator, and in the end, the latter tries to extract as much personal and confidential information as they can about their victim. In fact, this is very much similar to what is known as a "social engineering attack".

11. *The Body Shaming*:
    In this kind of scenario, the perpetrator makes very harsh and rude remarks about how their victim looks in terms of body shape, weight, image, and even size. A subvariant of this is known as "Fat Shaming", where obese victims are primarily targeted.

12. *The Swatting*:
    In this instance, Law Enforcement and other types and kinds of Emergency Personnel are sent to the home of the victim. It is the perpetrator that makes these kinds of calls, thus leaving the victim to be at blame and be held accountable for.

1.7 The Impacts of Cyberbullying by Category

In the last section of this chapter, we reviewed the major variants of Cyberbullying, and the venues in which they can be launched, which is namely the social media platforms. However, in this section, we take a different angle on Cyberbullying and review how its impacts are felt upon the different venues and age groups of the victims. This section will primarily focus upon:

- Cyberbullying and kids.
- Cyberbullying and adults.
- Cyberbullying and the workplace.
- Cyberbullying and the social media platforms.

1.7.1 Cyberbullying and Kids

As bad as it sounds, it is quite unfortunately the case that children are probably among the most which are Cyberbullied. There are a number of reasons for this, it could be the case that they are not brought up in a loving, family environment, and because of that, they suffer from a lack of self-esteem and even depression. Or it simply might be the case that they are being prone for getting picked on, whether it is in the classroom or even on the playground.

But in today's times, the biggest culprit of Cyberbullying against children is that they have easy access to a wireless device, even when their parents are not watching them. To prove just how bad it is, consider these statistics in the next subsection.

1.7.2 *The Statistics of Cyberbullying on Kids*

1.7.2.1 *Statistics by Country*

- 34% of British youngsters have experienced bullying in mobile games.
- In the country of Myanmar, any victims of Cyberbullying were mercilessly teased for having trouble concentrating on their school work.
- Kids in Japan between the ages of 12 and 18 are more likely to experience future health and social problems if they are the victims of Cyberbullying.
- In Russia, motive for Cyberbullying kids is sexual orientation, internet activity, or physical appearance.
- Foreign-born youths in wealthy countries experience more bullying than those who were born locally.
- In the United Kingdom, 17% of all adults report that their kids are being Cyberbullied in school.
- 33% of all young children experience Cyberbullying, on a global basis.
- 70% of young children with a physical disability are Cyberbullied.
- Young girls are 1.3 times more likely to be a victim of Cyberbullying than boys of the same age.
- 49% of LGBTQ+ of young kids have experienced Cyberbullying on a global basis.
- Overall, 18% of young kids report being a victim of Cyberbullying in the United States.

1.7.2.2 *Statistics by Social Media and Bullying*

- 31% of kids report that their classmates misunderstand their texts or social media posts on a regular basis.
- Kids in Romania are more likely to experience Cyberbullying if they use social media platforms for two or more hours per day.

- 9% of kids report malicious pictures of them being posted on social media sites without permission.
- Children ages 9–10 are more likely to be bullied on gaming websites, while kids from 13–16 are highly likely to be affected by Cyberbullying on the social media platforms.
- 42% of kids have experienced Cyberbullying on Instagram and 37% on Facebook.
- Overall, 38% of young kids see instances of Cyberbullying on the social media platforms on a daily basis.

1.7.2.3 *Statistics by Cyberbullying in the Schools*
- 9% of students between the ages of 12 and 17 admit that they have claimed to be someone else online.
- Cyberbullying is the #1 concern for school teachers.
- 25% of students report having skipped school because they were a victim of Cyberbullying.
- Cyberbullying happens the most in sixth grade, which makes up for 29% of Cyberbullying incidents.

(SOURCE FOR THE ABOVE STATISTICS: https://www.pandasecurity.com/en/mediacenter/ family-safety/cyberbullying-statistics/).

1.7.2.4 *How to Mitigate Cyberbullying in the Schools*
Like cyberattacks, Cyberbullying can really never be prevented 100%. But what parents can do is to mitigate the risk of that happening to their kid. Look closely at these Top 10 Tips:

1. *Apply Enforceable Boundaries*:
   Prevent altogether your kids from using either wireless devices and/or smartphones until you feel that they are ready and mature enough to do so. At this vulnerable age, young kids need to understand there is a difference between their physical identity and their digital identity.
2. *Maintain an Open Channel of Communication for Your Kids*:
   Always encourage your young kids to come to you with any questions about relationships at school and/or activity online.

3. *Have a Deeper Level of Conversation*:
   Whenever the time is appropriate, always discuss real life stories about Cyberbullying and other forms of online risks with your whole family. Use this as a venue for your kids to express themselves as to how they would react and respond to certain incidents of Cyberbullying.

4. *Mental Illness Is a Real*:
   As it was discussed before, young with depression and anxiety are often the prime targets for a perpetrator. If your kid is with depression and anxiety, make sure they get the right kind of help and that getting this does not make them different from anybody else.

5. *Watch for Behavioral Changes*:
   Young kids often exhibit signs of isolation, withdrawal, and aversion to activities around them. This is usually the first red flag that they have become a victim of Cyberbullying.

6. *Watch Out for Signs of Excessive Time Being Spent Online, or With Their Personal Devices*:
   If you see that your young kid has an uptick in online activity, or note that your kid seems all of a sudden with their wireless device, this is also *a huge warning sign*. Always be sure to check your online account and immediately discuss your concerns with them.

7. *Be Calm, Compassionate, and Understanding*:
   If your child tells you that they have become a victim of Cyber, *do not scold or punish them!!!* It is very crucial that you work together as a team in order to decide what the next actions should be.

8. *Ask Them Directly What They Want*:
   If your young child has become a victim of Cyberbullying situation ask them directly about the outcome that they want to see happen, and from there, work with them to come up with the appropriate solution.

1.7.3 Cyberbullying and Adults

Just like kids, adults can also become quite easily the victim of a Cyberbullying attack. Consider some of these statistics as outlined in the next subsection:

1.7.3.1 Statistics by Adult

- 64% of American adults (from 18 to 29) have become the victim of Cyberbullying in recent years.
- Young adults of Cyberbullying are almost twice as likely to attempt suicide than others.
- 41% of all adults in the United States have been victims of online harassment.
- The total number of American adults experiencing physical threats and sexual harassment has doubled since 2014.
- 75% of the victims were harassed on Facebook.
- 92% of all American adults think that Cyberbullying is a national problem.
- 41% of adults worldwide who have received physical threats on the social media platforms claim that the particular platform did not take any kind of action.
- 79% of all US adults don't think that the social media platforms do a good enough job at addressing the issues of Cyberbullying.
- When adults globally have been asked what would be the most effective response to Cyberbullying, 51% strongly believed that there should be permanent bans for offending user's social media accounts.
- 80% of adult Americans want more Federal Laws to address Cyberbullying.
- Women adults are more than twice as likely than men adults to find the effects of Cyberbullying to be very upsetting.
- Approximately 25% of adults who have become Cyberbullying targets reported having trouble sleeping at night.
- 16% of adults who were a victim of Cyberbullying took the extreme steps such as moving to a new location or taking self-defense classes.
- Political views and beliefs are the main reason why adults become a Cyberbullying target.
- 9% of adult victims said they were bullied because of their political views, followed by physical appearance at 36%, then gender at 28%, trailed by ethnicity and religion at 28% and 21%, respectively.
- A recent Good Housekeeping survey found that 89% of adults have been "Fat Shamed".

- 75% of US adults who have been Cyberbullied have experienced it on the social media platforms.
- 25% of all adults say that their latest incident of Cyberbullying was on an online forum or discussion site; 24% say that they were a victim of Cyberbullying through texting or a social media app.

**(SOURCE: https://explodingtopics.
com/blog/cyberbullying-stats).**

1.7.3.2 How to Mitigate Cyberbullying in Adults
As we have seen in this chapter, anybody can become a victim of Cyberbullying. It does not matter the age, social status, ethnic background, socioeconomic status, geographic location, etc. Kids all the way to the oldest of the senior citizens can become a victim. Worst yet, you could still be Cyberbullied, and not even know your antagonist.

So what can somebody, especially an adult, do in this case? Here are some tips to follow:

1. *Recognize What Cyberbullying Really Is:*
   It is very important to recognize what Cyberbullying is and not avoid the fact if you have become a victim of an attack. Cyberbullying is clearly defined at the beginning of this chapter.
2. *Keep Careful Documentation:*
   By keeping detailed records of the Cyberbullying attacks against you, this can make it easier to report them to the respective social media platforms and/or Law Enforcement.
3. *Use Built-In Blocking, Ignoring, and Reporting Tools:*
   Almost every social media platform online has the option to block, ignore, and report perpetrators who engage in threatening behavior and further behaviors of aggression. Use these tools and functionalities to your fullest advantage.so that you can protect yourself. But some antagonists are very adept at getting around these blocking mechanisms.
4. *Get Law Enforcement Involved:*
   Contact Law Enforcement to find out if anything can be done about your Cyberbullying problems. This is especially

important if anyone online threatens you physically and/or your family.

5. *Don't Retaliate*:
   It is important not to respond to the perpetrator except to calmly tell them to stop. If you fight back, it will only get worse and much more confrontational.

1.7.4 Cyberbullying and the Workplace

Ever since the COVID-19 pandemic, most of the workforce in Corporate America has now become remote based. From the standpoint of Cybersecurity, there are many issues, especially that of heisting of video conference meetings (one such breach was called "Zoombombing") and the dangers that lurked because of intermeshing with the unsecured home networks with that of the fortified corporate networks. But apart from this, another phenomenon happened: the uptick in Cyberbullying to these kinds of worker. We address this in this subsection.

1.7.4.1 Statistics by the Workforce

- From 2017 to 2022, 31% of the American workforce experienced some kind of Cyberbullying.
- The United States has a workforce of about 157 million people, and 31% of them have experienced some sort of Cyberbullying.
- In terms of the remote workers, 61.5% have reported being Cyberbullied.
- Most of the Cyberbullying occurs during online meetings (50%).
- Most of the Cyberbullying at the workplace takes place via email (9%).
- 65% of all Cyberbullying for remote workers is top-down in the United States.
- 14% of the Cyberbullying is bottom-up, launched by other coworkers.
- Male workers were Cyberbullied at 67%, while women made up the remaining 33% of total victims.
- 72% of Cyberbullying in the workplace were lone acts, while 28% involved more than one perpetrator.

- The most common age group to be bullied in the American workforce is between 17 and 29.
- In terms of other ethnicities, 35% of Hispanic workers and 26% of African Americans have been victims of Cyberbullying.
- In 67% of all Cyberbullying incidents, the victim of Cyberbullying ends up leaving their current job.
- In 20% of the cases, the perpetrator is terminated from their job.
- 75% of employees who have been a victim of Cyberbullying have reported psychological distress.
- About 55% of the perpetrators themselves have been targets of Cyberbullying.
- Social media platforms are still a favored venue for Cyberbullying, with 45% of employees being targeted here.
- 60% of the workplace victims also suffer from deep levels of anxiety.
- 28% of the workplace victims were targeted because of discrimination based on race or ethnicity.
- Around 21% of employees are targeted because of their sexual orientation.
- Approximately 40% of Cyberbullying victims at the workplace experience emotional stress that interferes with their job performance.
- 35% of employees are Cyberbullied through SMS-based text messages.
- 50% of the workplace victims do not know about any procedures or protocols that are currently in place for reporting Cyberbullying.
- Only about 14% of Cyberbullying cases ever come to any kind of final resolution.
- 80% of all Cyberbullying incidents in the workplace are of a personal nature.
- 24% of employees in the American workforce have witnessed other coworkers being directly Cyberbullied.

(SOURCES: https://www.thehrdirector.com/ features/diversity-and-equality/counter cyberbullying-workplace/https://zipdo.co/ statistics/cyberbullying-in-the-workplace/).

1.7.4.2 How to Mitigate Cyberbullying at the Workplace

Cyberbullying at the workplace is something that should be taken very seriously. After all, in these kinds of instances, the employee, if he or she believes is being bullied, can easily file a lawsuit against the business. Likewise, if the employee is Cyberbullying against another entity, and if they are using company-issued devices, the business owner can also be subject to a lawsuit.

It is important to keep in mind that while Cyberbullying should never be tolerated no matter what the venue or medium is, it can have the most severe legal consequences and repercussions in the workplace. Because of this, managers need to be proactive about this, but in the end, the "buck" literally has to stop with the Human Resources (HR) department for not only making sure that Cyberbullying against employees within the confines or even outside of it as well, but they also need to make sure that any type of punishment to the perpetrator must be strictly enforced.

So, what can an HR department do in these situations? Here are some ideas:

1. *Understand What Cyberbullying Is All About:*
   HR professionals need to know the differences between Cyberbullying and other types of work-related complaints, especially that of "Traditional Bullying", as detailed earlier in this chapter. For example, it can happen through phishing-based emails, text or SMS messages, and once again the social media platforms. In these kinds of instances, passive-aggressive behavior by the perpetrator is the most common behavior that can take place.

2. *Launch a Formal Investigation:*
   The HR team has a legal obligation to investigate all reports and cases of Cyberbullying that takes place in the workplace. For example, it may be necessary to offer the bullied employee different mechanisms to manage any stress, anxiety, or depression as a result of the incident. In regard to the perpetrator, that particular person may not realize that what they have done. As a result, coaching or sensitivity training may be needed in these cases.

3. *Contact Law Enforcement:*
   The HR department must immediately notify Law Enforcement for any reported cases of Cyberbullying in the

workplace. This will also serve as a protective mechanism for the company in the case a lawsuit is filed against them.

4. *Receiving a Formal Complaint*:
   As a member of HR you should need to take immediate action by speaking with the victim and the perpetrator, on a confidential basis. These conversations must be documented and recorded and stored for a predetermined time period.

5. *Be Fair*:
   You must be objective and fair in your investigation of the Cyberbullying incident. The investigators must protect the privacy of everyone involved in the case. Then, the appropriate follow-through must be done to enforce disciplinary action and restore order among all of the employees.

Other general tips the HR must also follow include:

- Establish an anti-Cyberbullying policy.
- Employees should receive training on Cyberbullying to recognize it and avoid it.
- Businesses must train their HR staff to fully investigate claims of Cyberbullying.
- Make the needed changes to the business and its work culture.
- Implement strategies to help strengthen individual managers and leaders.

1.7.4.3 The Financial Consequences of Cyberbullying at the Workplace
Apart from the mental and legal ramifications of Cyberbullying in the workplace, there are also severe ones to the employer. These are the ones that impact the bottom line. Here are some examples of the financial toll Cyberbullying can take on a business:

1. *Loss of Productivity*:
   Employees don't usually do well in high stress situations, especially when Cyberbullying is involved. For example, the most common symptom of this is a sheer lack of motivation. This, in turn, leads to a marked decrease in productivity.

2. *Healthcare Costs*:
   Cyberbullying can have a horrible impact on the physical health of the employees, such as high blood pressure,

depression, migraine headaches, or anxiety. This can cost an employer in the form of sick leaves and rising health insurance costs.

Also, if an employee chooses not to quit their job, the employer may be forced to pay both medical and mental health costs to help the employee improve their well-being.

3. *Absenteeism*:
When employees are Cyberbullied, they are much more likely to be sick when they're not, or may even take extended medical leaves. This results in big loss of time in the workplace.

4. *Workplace Turnover:*
Cyberbullying in the workplace has been often correlated with a high employee turnover rate. There are costs that are associated with recruiting, hiring, and training their replacements.

5. *Reputation Damage:*
The company can also experience what is known as "Reputational Damage". This can lead to hiring difficulties. Also the business in question will have difficulty selling its product and services. This is even further exacerbated by online reviews that can be quickly posted on Yelp, Zillow, Google, etc.

1.7.5 *Cyberbullying and Social Media Platforms*

As it has been reviewed throughout this entire chapter, it is the social media platforms that are the biggest culprits for Cyberbullying, no matter what the particular environment is. But in order to paint a true picture of it, and to further drive home the sheer gravity of it, we devote the next subsection to further examining even more detailed statistics.

1.7.5.1 *Statistics by the Social Media Platforms*

- Of the total number of victims on the social media platforms, most have been middle and high school students. Of these, 37% have experienced online harassment of some variant.
- 90% of teens agree that Cyberbullying on the social media platforms is a major problem that has a negative impact on their relationships.

- 41% of American adults have faced Cyberbullying on the social media platforms, and also 67%of them view it as a major problem in society as a whole.
- The age group of 18 − 29 are the prime targets for Cyberbullying on the social media platforms, and in fact, 41% have experienced extreme forms of it.
- 9% of adults aged 30 − 49 have experienced online harassment.
- In the age group of 50 and above, the victims have reported that they have been targets of hostile behaviors on the social media platforms. Of these, 44% are male and 37% are women.
- Women in the age group of 18 − 29 reported experiencing online sexual threats, and only 9% of men have had the same experience.
- 60% of both teen boys and girls have been targets of harassment on the social media platforms.
- 42% of all teenagers reported that they have been called by offensive and ridiculing names on the social media platforms.
- 32% of victims said they have been the targets of false information, such as Deepfakes.
- 25% of victims on the social media platforms have received pornographic images that they did ask for.
- 22% of victims have experienced "online shaming".
- 8% of the Cyberbullying victims have received threats of physical harm on the social media platforms, whereas 8% have actually experienced online stalking.
- 7% of the victims claimed that they were harassed continuously on the social media platforms, and 6% of them were of a sexual nature.
- So far, Facebook is the biggest offender on the social media platforms, with 77% of victims being directly Cyberbullied on it.
- Twitter was at 27% for Cyberbullying.
- YouTube and Instagram, followed at third place for Cyberbullying, with 18% and 17% of total reported cases, respectively.
- The 18 − 49 age bracket is more likely to be Cyberbullied on the social media platforms while adults in the ages 50 and

older age bracket experience more Cyberbullying via email messages.

(SOURCE: https://research.com/education/ teenage-cyberbullying-statistics).

1.7.5.2 The Other Social Media Platforms That Are Venues for Cyberbullying
It is important to note at this point that it is not just the major social media platforms such as Facebook, LinkedIn, X (formerly "Twitter"), Instagram, Pinterest, and YouTube that are the agents for Cyberbullying. There are many others as well, which include the following:

- Amino
- BeReal
- Discord
- Reddit
- Telegram
- TikTok
- WeChat
- WhatsApp

The following are the platforms that allow users to share videos and receive comments, subscriptions, and follows from other users:

- YUBO
- Askfm
- Calculator%
- Chatroulette
- Discord
- Houseparty
- Kik
- Line
- LiveMe
- MeetMe
- Omegle
- Roblox
- Sarahah

- Snapchat
- Telegram
- Tumblr
- Twitch
- **VSCO**
- WeChat
- Whisper
- YUBO (formerly YELLOW)
- YouNow

1.8 Prelude to Generative AI

In this chapter, we have provided an extensive review into what Cyberbullying is all about. The rest of the book will now focus on Generative AI, and how it can even be possibly used to help mitigate Cyberbullying.

In today's society, we now live in a world where artificial intelligence (AI) and machine learning (ML) are making the news headlines on an almost daily basis. While these concepts have been around since the 1950s, its recent eruption has been brought primarily by ChatGPT and the GPT4 algorithms that drive it. AI and ML can be used for both good and bad. For example, when it comes to Cybersecurity, these tools can be used to predict future threat variants in a matter of minutes where it would take a human being days to figure this out.

But, on the flip side, AI and ML can also be used by the Cyberattacker to launch even more nefarious kinds of malicious payloads, especially when it comes to ransomware. They can now make them much more covert in nature, which makes it even that much harder to detect in time. In fact, it is the Generative AI that is primarily used to launch Cyberbullying attacks on the various social media platforms.

Both AI and ML tools have made it easier for trolls to seek out vulnerable targets and create convincing fake content, of all types. One example of this is Deepfakes. With this, trolls can create fake videos or images of real people, which are difficult to detect at first glance. As a result, this makes it very easy to spread false information or damage someone's reputation. Chatbots, although they have been around for quite some time, are also being used to flood social media platforms with abusive messages and also spread false news.

To make matters even worse, Generative AI can also be used to create real messages to the victim in order to lure them into a trap. It is important to note that Deepfakes were reviewed in detail earlier in this chapter, but a troll can be technically defined as follows:

> A Troll is a term for a person, usually anonymous, who deliberately starts an argument or posts inflammatory or aggressive comments with the aim of provoking either an individual or a group into reacting.
>
> **(SOURCE: https://bulliesout.com/ need-support/young-people/trolling/)**

But Generative AI can also be used in mobile apps that are designed to combat Cyberbullying. This is reviewed in the next chapter.

Chapter 1 Resources

1. https://www.dispartilaw.com/what-is-cyberbullying-definition/#:~: text=Traditional%20bullying%20typically%20refers%20to,physical%20 aggression
2. https://kidshealth.org/en/teens/cyberbullying.html#:~:text=Cyber bullying%20is%20the%20use%20of,hurt%20or%20embarrass%20 someone%20else
3. https://www.adobe.com/express/learn/blog/what-is-a-meme
4. https://www.pandasecurity.com/en/mediacenter/family-safety/cyber bullying-statistics/
5. https://explodingtopics.com/blog/cyberbullying-stats
6. https://www.thehrdirector.com/features/diversity-and-equality/counter-cyberbullying-workplace
7. https://zipdo.co/statistics/cyberbullying-in-the-workplace
8. https://research.com/education/teenage-cyberbullying-statistics
9. https://bulliesout.com/need-support/young-people/trolling/

2

THE USE OF TECHNOLOGY TO COMBAT CYBERBULLYING

In the last chapter of this book, we provided an extensive overview into what Cyberbullying is all about. We first started with an examination of what "Traditional Bullying" is, and from there, there was an extensive review as to how Cyberbullying actually became a derivative of it. From there, we then explored as to how technology, especially Generative AI, can be used as a tool for not only launching Cyberbullying attacks, but as to how it can be used to prevent them as well. This will be explored in further detail in the next chapter.

But before we actually do a deep dive into Generative AI, it is first important to note the tools that can be used to actually fend off the Cyberbullying attacks. There was some focus that was paid attention to mobile apps in the last chapter as well. Given the extensive usage of them, mobile apps have now become a highly favored target to not only find their particular victims but also launch their attacks against them.

But just like Generative AI, there is also a flip side to this. For example, mobile apps can also be used to combat Cyberbullying attacks. It can be used by individuals by themselves, but is more widely used by teachers and parents. Also, Generative AI can be used in the creation of the mobile apps, in order to make them "smarter" and stay one step ahead of the perpetrator.

The usage of these kinds of mobile apps is now reviewed in more detail in the next section.

2.1 The Top Anti-Cyberbullying Apps

There are a plethora of mobile apps that parents can use to keep tabs on what their kids are doing, especially when it comes to Cyberbullying. In this section, we review some of the major ones:

36   DOI: 10.1201/9781032666020-2

1. *The Net Nanny*:

   "Net Nanny" is deemed to be one of the first mobile apps to have come out onto the marketplace. It can be used to track all kinds and types of communications that your kids may have on the social media platforms by also making use of what is known as the "Net Nanny Social Ad". Parents have full control for blocking and/or restricting messages.

2. *The PhoneSheriff*:

   With the "PhoneSheriff", parents can on a real-time basis closely examine what their kids are doing, sharing, or communicating with others. This can be installed on any device, such as the iOS- and Android-based ones. Also, additional filtering options are integrated into it, and as a result, parents can select the right kinds and types of activities that they want their kids to be engaged in. Additionally, parents have the ability to blacklist websites, and it even comes with a GPS tracker in order to track the perpetrator.

3. *The Mobile-Spy*:

   This kind of mobile app permits the parents to monitor their child's online activities in real time by accessing all of their messages. Snapshots are also provided as to what your child is doing on any particular device. This can also be used as evidence for Law Enforcement in order to bring the perpetrator to justice.

4. *The Qstudio*:

   This is a specialized kind of mobile app that can be easily installed on any type or kind of wireless device, whether it is iOS or Android based. Best of all, it has a content filtering mechanism as well and will notify parents if there is any malicious activity that is taking place. Another powerful functionality of this mobile app is that it will alert the parent of any new contacts or friends that your child has made in their social media platform browsing activities. Even better, the emergency-based functionality will notify the parent of their child's exact and precise geo-location at any given moment that it is requested.

5. *The Re-Think*:

   This is one of the most famous Anti-Cyberbullying apps available today. In fact, it was created by a teenage girl who was an actual victim of Cyberbullying. This app is deemed innovative and non-intrusive to both the parents and the kids. The primary goal here is to prevent Cyberbullying incidents even before it starts. Also, this particular tool red flags messages that are hateful or abusive.

6. *The Bully Block*:

This is actually a free mobile app that is designed for kids to deal with Cyberbullying when it actually happens on a real-time basis. Parents can also covertly record and capture any kind or type of Cyberbullying incident as it happens on a real-time basis and use it as evidence to Law Enforcement. Even more, parents can easily record the perpetrator and hold responsibility for the act.

7. *The Managed Methods*:

This is a Cloud-based security and monitoring platform designed exclusively for teachers to help protect their students. It makes use of API integrations in conjunction with Microsoft 365, which are the main software tools used by school districts nationwide. Confidential and private information/data about your students are always safeguarded.

This mobile app can alert both teachers and parents in just a matter of minutes of any Cyberbullying attempts on the students. Also, you can receive the most comprehensive details about the Cyberbullying attempt.

8. *The Toot Toot*:

This mobile app was developed in partnership with Barclays. It allows students to report concerns about all forms of Cyberbullying.

9. *The Find My Kids – Footprints*:

This mobile app is exclusively made available for the iPhone. With this, a parent can track their child's movement and set up the geo-location functionality.

10. *The Sit With Us*:

This mobile app was created by a 16-year-old girl who was once a victim of severe bullying during her school days. The platform allows students to set up inclusive lunches with their classmates. Their motto: "No One Needs To Eat Alone!"

Kids can also become "Sit With Us Ambassadors", which serves as a confidence boost too.

11. *The Sarahah*:

This kind of mobile app was initially designed to be a workplace productivity tool, but soon it became a vehicle that teenagers used for massive levels of Cyberbullying. Because of this, both Apple and Google banned it from their respective stores. Ironically, "Sarahah" means "Honesty" in Arabic. Once this particular mobile app has been procured by a business, the appropriate manager will receive a specialized code and a link to give out to employees.

To help combat Cyberbullying in the workplace, any messages that are transmitted are sent as direct one, thus allowing representatives from the company to follow up with the suspected employee if they need more information and take next steps for disciplinary action.

2.2 The Use of Parental Monitoring Mobile Apps

In order to control the effects of Cyberbullying on your kids, the best and probably the most drastic course of action you can take is to simply not allow them to use any kind of wireless devices, until you feel that as a parent, they can use such things in a responsible fashion. But the bottom line is that smartphones and other related devices are within easy reach of everybody.

So, if you have strict rules about not using them, your kid quite easily retaliate and find a secretive way to get access to one, without your knowledge. In the end becomes a careful balancing act, and therefore, a good solution to use here would be what is known as a "parental monitoring app". These are also mobile downloads, but the benefit here is that while your kid still may have access to a wireless device, you will still be able to see what they are up to. This will then give you the control to stop any suspicious activity before it actually happens.

Here are some of the primary benefits of using a parental monitoring app:

• You can track and monitor SMS, calls, and emails on a real-time basis.

- You can be alerted for any suicidal indications and Cyberbullying immediately.
- This can contribute to the development and teaching of Cyberbullying self-defense.
- You can quickly and easily block and filter any unwanted applications and websites.
- You can set time limits for using any social media platform, such as Facebook, X, Instagram, and Pinterest.

2.3 The Top Parental Monitoring Mobile Apps

Some of the best tools that you can use in this regard include the following:

1. *The Net Cut*:
   More information about this can be found at the following link:

   https://arcai.com/netcut/

2. *The Family Time*:
   More information about this can be found at the following link:

   https://familytime.io/?clickid=2fb0zC1flxyPTkATI-R9OzQYUkHwzMV1QRfjXQ0

3. *The DNS Filter*:
   More information about this can be found at the following link:

   https://www.dnsfilter.com/

2.4 The Other Ways That Anti-Cyberbullying Apps Can Be Used

In this section, we take a more macro approach as to how technology in general can be used to help stop Cyberbullying. To get started with, here are some tips on how to best use your mobile app of choice:

- Make sure the mobile app runs on multiple types of devices. For example, if your parenting app is on iPhone, make sure that it can work on an Android-based device, such as the Samsung smartphone.

- Make sure your school district has an acceptable device use policy. Students, parents, and teachers need to agree to this, and in writing. This policy should include how students should be informed about how their online activities will be monitored.
- Make sure the parental monitoring app program doesn't just monitor the major social media platforms. It is important to keep in mind that there are plenty of other sites that students can use for launching Cyberbullying attacks.
- Make sure that your parental monitoring app provides an anonymous reporting feature so that the friends of your kids have another avenue to report concerns.
- Give the idea for your school districts to also teach "digital citizenship lessons". This will help students learn how to use the social media platforms in a safe manner. This will help to keep students safe online.
- Use any type or kind of incident as a learning opportunity, as Cyberbulliers know that their actions have blasting effects on their victims, and they need to have reassurances that they are being helped.
- Maintain both Cyberbullying Reporting Hotlines and monitor Contact Form Submissions on a 24 × 7 × 365 basis. These kinds of Hotlines, whether they're accessible by phone or online, are a great reporting tool for victims or witnesses of incidents. Parents, teachers, and kids can remain anonymous, thus eliminating the fear of retaliation and torment.
- Incident Tracking and other types and kinds of Case Management Software Packageshelp school administrators to respond to any Cyberbullying issues before they explode into major ones. Also, the data that is generated will help to mitigate any future issues. This type of software provides an overview of all complaints and investigations that have been logged into the system, thus allowing better oversight and risk management on part of the parents, teachers, and other school staff. This helps to create evidence-based approaches that can also be submitted to Law Enforcement.

2.5 The Startups Leading the Way in Innovations Using Generative AI

There is no doubt that the Generative AI startups (also known as "SMBs") are the fuel to the American Economic Engine at the present time. In this section, we review a sampling of them:

1. *Bodyguard*:

   This startup offers a mobile app that protects individuals and businesses against Cyberbullying on social media. Through comprehending text, it acts as a real-time agent, monitoring all types and kinds of content between kids. If anything is detected, the conversation is then deleted before anybody is hurt. Equally important, the app also notifies parents via email if their kids receive a comment from a bully, or are the creators of hateful comments.

   You can view the website at the following link:
   https://www.bodyguard.ai/

2. *Tall Poppy*:

   This entity is a US-based startup that helps companies to protect their employees from Cyberbullying. This mobile app includes password management, access to legal resources, etc. The impacted employee can reach out to an Incident Responders to prepare them to work effectively with Law Enforcement. Another added benefit is that the platform also provides anti-harassment training to the employees.

   You can view the website at the following link:
   https://www.tallpoppy.com/

3. *Kidas*:

   This entity is an Israeli-based startup that interestingly enough makes Anti-Cyberbullying software for online video games. The app automatically scans all forms of communications and alerts parents through emails when it detects Cyberbullying or other online predators.

   You can view the website at the following link:
   https://getkidas.com/

2.6 The Creation of Anti-Cyberbullying Mobile Apps

There is no doubt, as it has been described in this chapter, the creation and deployment of mobile apps will continue to grow at an exponential rate, no matter what the industry they serve. But these days, Generative AI is now also being heavily used to not only create the source code of these mobile apps but also serve as a double check to make sure that any gaps, weaknesses, or holes are uncovered. From here, it is then of course up to the CISO and the IT Security to make sure that all of the Controls have been put into place to correct them, and that they also work as well.

Since this chapter has covered the usage of mobile apps as it relates to helping combat Cyberbullying, it is important to give a technical understanding of these applications in three key areas:

- The top mistakes that are made by software developers.
- How these mistakes can be rectified.
- The role that Cybersecurity plays in with the mobile apps.

2.6.1 *The Top Mistakes That Are Made by Software Developers*

1. *Not Meeting the Needs of the Parent*:
   When it comes to Cyberbullying, you need to meet the needs of the parents and the school educators. After all, it is the kids who need protection. So, this is the time that you really need to sit down to understand what they really need. This can be one of the hardest and longest phases of the mobile app development lifecycle. Also, as the mobile app is being developed, make use of what are known as "Mockups", and take a modular approach to make sure that all of the needs and details are met. Parents and school educators are not enamored with bells and whistles; they just want a mobile app that will work.

2. *Creating for Too Many Platforms*:
   The top three sources for mobile app downloads are Google, Apple, and even Microsoft. While your ultimate goal is to have your mobile app on every store that is possible for the parents and school educators to download, you need to take a slow and cautious approach to this. If your intention is to create mobile apps for all of the school districts, then your goal should be to make it available across all of these stores. If not, then you need to take a much focused approach to determining who will get this mobile app. This also

serves a Cybersecurity purpose. By taking a much slower approach, your mobile app will not catch the direct eye of the Cyberattacker.

3. *Not Having Enough Tools to Help*:

   In this regard, you want the parent and school educators to understand how to use your mobile app quickly. In this regard, there is absolutely no need to write a 100-page manual, but rather, keep it concise filled with visuals, as your customer goes through it, filled with visuals. Also, from the standpoint of User Interface (UI)/User Experience (UX), create a simple and clean design. Even more important, that the content is appropriate to the culture of the people that will be making use of the mobile app.

4. *Overloading the App*:

   In this area, give exactly what the parents and school educators need. Once the mobile app has been accepted and used by them, then consider adding more upgrades to it over time. Don't make the mobile app "bloated", if not could very well break, and even cause a security issue.

2.6.2 How These Mistakes Can Be Rectified

The following are a set of guidelines that can be used to correct and remediate the mistakes as reviewed in the last subsection:

1. *Keep the Navigation Easy*:

   It is in the mindset of most software developers that by making the mobile app something more complex, it will have a broader appeal to the parents and school educators, because you are showing off what you can do. But this is far from the truth. They need and want something that is easy to navigate, and track any type or kind of harassing messages to the kids on a real-time basis. If not, they will quickly abandon your mobile app in search of something else.

2. *Choose the Right Programming Language or Script to Create the App*:

   With this in mind, you need to select the appropriate programming language and/or scripting language to create the mobile app in the most efficient and most secure way possible. Here is what you can make use of in this regard:

- Ruby On Rails
- Python
- Perl
- C
- C++
- HTML (also known as "Hyper Text Markup Language")
- Java
- SQL (also known as "Structured Query Language")
- PHP (also known as "Hypertext Preprocessor")

3. *The Thumb Rule*:

This is a very crucial area not just from the standpoint of the UI/UX process, but also in terms of access by the parents and school educators. This particular phrase simply means that your mobile app should be easily procured and utilized from the perspective of using the thumb only. As an analogy to this, think of when you make use of your smartphone next time: For the most part, you are making use of your thumb to navigate through both on the keyboard and on the screen everything. Your mobile app should also be created in this very same manner, so that it is both thumb scrolling friendly and thumb interactive friendly.

4. *Always Keep Up With the Testing*:

When creating your mobile app, it is extremely important to keep testing your mobile app, and this process should be done over and over again. This is to ensure that to make sure not only that you have a secure product, but that all gaps and vulnerabilities have been found and quickly remediated. Typical tests that you can use for your mobile app include the following:

- UI/UX Testing (to make sure that the environment of the mobile app is easy to use)
- Memory Leak Testing (this is to ensure that any information and data that are stored in the mobile app are mitigated as much as possible from any chances of a Data Exfiltration Attack).

- Platform and Device Testing (this is to ensure that your mobile app not only works but is also secure from the standpoint of interoperability with the major operating systems, primarily that of the iOS and the Android platforms).

5. *Break the Process Down into Smaller Steps*:
   It is very important to keep in mind that the creation and development of a mobile app can be quite complex, depending upon the needs and requirements of parents and the school educators. Therefore, you must make strong usage of the principles of Project Management and break down actions into smaller action items that are much more digestible for you to follow. In other words, you should never take the approach of creating a mobile app by doing it all at once, like chopping down an entire tree in one full swipe all at once. In the world of both mobile app and software development, this is known as taking a "Modular Approach". Put in simpler terms, all of the source codes that are used to create the mobile app are broken down into separate "Modules" or components. By taking this kind of particular approach, not only will it be easier and much more efficient to test the Quality Control (also known as the "QC"), but it is also far more robust to test it also from the standpoint of Cybersecurity.

Apart from the issues just reviewed in the last two subsections, Cybersecurity is very often an ignored issue when creating and developing a mobile app. Here are some areas that get overlooked, and thus can put the child at a far graver risk:

1. *Encryption Not Being Used*:
   Simply put, this is scrambling of information and data at the point of origination and unscrambling the data at the point of destination. The main goal is to keep the information and data in a garbled state that it will be rendered useless if it were to be intercepted by a malicious Third Party. It is not just mobile app developers that do not take this into consideration, but software developers in general. If their personal information and the child data are not encrypted (primarily the credit

card number and social security number), they could easily become the victim of ID theft or other threat variants. In these cases, all that is needed by the Cyberattacker is a basic Network Sniffer to capture the Data Packets coming from the wireless device. This is also known as a "Man In The Middle Attack".

2. *Outsourcing the Mobile App Development Process*:
    This comes down to the basic matter of whether to outsource your mobile app development or not. In this regard, the primary consideration is that of cost. It is important to keep in mind that the mobile app is also a software development process at its structure. The bottom line here is that you get what you pay for, and thus in this regard, it may be far wiser to get an experienced development team involved in the entire process.

3. *Keeping a Unified Structure*:
    With this, it is imperative that you stick with and enforce a consistent formatting standard and best practices, both for the end user and administrative documentation, and any comments that you may leave in the source code of the mobile app for future usage. In terms of the latter, creating comments is a very important component to have, especially when it comes to time to come out with later upgrades to the mobile app. This will help in creating the structure also for the source code. Thus having source code that this written and also compiled a clean, organized structure will make it efficient and effective to go back and correct, if necessary.

4. *Not Testing the Code*:
    This is by far the most overlooked area in all of software development. The source code rarely ever gets tested, and if does, it is usually done at the very end, right before delivery to the client. According to the latest Verizon Security Breach report, mobile app security is becoming a prime concern today. Therefore in this regard, both Vulnerability Scanning and Penetration Testing to the source code must be done at the most granular level possible. Quality Assurance (QA) testing needs to be done as well. Therefore, it is highly

advised that both types of testing take place at a modular level, so that any flaws do not have a cascading effect in other parts of the mobile app development process.

5. *Making Use of APIs*:

   In all kinds and types of software development that takes place today, software developers often like to use what are known as "APIs" (which stands for "Application Protocol Interface"). These are simply lines of code that have been created in a standardized format that can be used to bridge the front end (the user interface) and the backends (the database) of any type or kind of mobile app. The beauty of this approach is that the source code that resides in the APIs can be used to modify to fit the needs and the requirements of the mobile app, thus ensuring a timely delivery of it to the client. This will obviously save a lot of time and money. But the main issue here is that software developers use APIs from open-source libraries that have not been kept up to date with the latest patches and upgrades. This of course can be a serious security flaw and can have detrimental and even graver consequences down the road if they are not checked before they are first used.

6. *The Need to Implement Other Protocols*:

   In this scenario, many mobile app developers fail to implement what are known as "Identity and Access Management" (also known as "IAM") protocols into the source code. This former refers to the 100% confirmation of the end user who is using the mobile app. In order to have a robust mobile app in terms of Cybersecurity, you should implement into it what is known as "Multifactor Authentication", or "MFA" in short. This is where at least three or more differing authentication mechanisms are used to confirm the identity of the user without a doubt. For example, these mechanisms could be a combination of passwords, challenge/answer responses, RSA tokens, and even biometric modalities (such as that of fingerprint recognition and/or iris recognition). Software developers need to factor this heavily at the very beginning stages of the mobile app creation, and make sure that it has been included into the source code at each and every point in time a new release of the mobile app actually comes out.

2.7 Gauging the Effectiveness of an Anti-Cyberbullying Mobile App

Once you have developed this kind of mobile app, and it is in usage by the parents and school faculty, you will want to get direct feedback as to how they view it. For example, was it useful to them to keep track of the child's online behavior? Did it mitigate any Cyberbullying attacks from actually happening? What advice do they have for any future functionalities of the mobile app?

These questions and many more can be answered by creating and delivering a survey instrument. The survey that we will propose in this section is based on the concepts of what is known as "Technology Acceptance Model", also known as "TAM" in short. But first, it is very important to give an overview of its theoretical concepts before introducing the actual survey.

2.8 The Definition of Technological Acceptance

At the very core of TAM, and as its name totally implies, is the term "Technology Acceptance". This can have different meanings to different people, but for purposes of this book, it can be defined as follows:

> It is based on the idea that our attitudes towards technology are shaped by two key factors: perceived usefulness and perceived ease of use. Perceived usefulness refers to the extent to which we believe that using a technology will enhance our performance or achieve our goals, while perceived ease of use refers to the degree to which we believe that using a technology will be effortless and straightforward.

> **(SOURCE: https://www.enablersofchange.com.au/ what-is-the-technology-acceptance-model/)**

There are a number of key components to this definition, which are as follows:

- There must be willingness by the end user to fully adopt the technology for use.

- The specific use must be of free will.
- The will to use the certain piece of technology cannot be forced upon to the end user.
- The technological device system must be adopted in its whole or entirety. Any partial usage of it is unacceptable.
- Acceptance is deemed to the "final make or break" for the technological device in question.
- It has also been reasoned that if the technological device is adopted in its full entirety by the will of free choice, then more usage will be demanded from it, as a result.

2.8.1 The Theoretical Components

It is important to note that the TAM did not just evolve off from its own. Rather, there are other theories that are related to Technological Acceptance that gave rise to TAM. Some of the major ones include the following:

1. *The Cognitive Dissonance Theory*:
   The Cognitive Dissonance Theory (CDT) is used to explain how discrepancies (dissonance) between one's mindful cognition and reality can change the person's subsequent cognition and/or behavior. It is important to note that major part of this theory comes from the end user's preset expectations of the IT system, assuming they know what particular system is about, to varying degrees.

2. *The Task Technology Fit Model*:
   The Task Technology Fit (TTF) model holds that an IT system is more likely to have a positive impact on individual performance and can be used if the capabilities of IT match the tasks that the user must perform. In other words, if the end user (such as the remote employee) feels that the system is helping them to a great extent in achieving their daily objectives, then the IT system will be accepted almost immediately, with no hesitance involved.

3. *The Theory of Reasoned Action*:
   The Theory of Reasoned Action (TRA) model has been deemed to be the first to gain widespread acceptance in technology acceptance research. It is a versatile behavioral theory and statistically models the attitude–behavior relationships.

4. *The Theory of Planned Behavior*:
   The Theory of Planned Behavior Model (TPM) is deemed to be a successor of the TRA model and it introduced a new variable known as the "Perceived Behavior Control", or "PBC" in short. It is determined by the availability of skills, resources, and opportunities, as well as the perceived importance of those skills, resources, and opportunities to gain specific outcomes.

5. *The Model of PC Utilization*:
   The underpinning paradigm in the Model of PC Utilization (MPCU) is the theory of human behavior. In other words, the model predicts the adoption and utilization of an IT system based on end user utilization behavior. The primary goal of this model makes it particularly suited to predict ultimate IT system acceptance and use across a wide range of IT systems.

6. *The Decomposed Theory of Planned Behavior*:
   The Decomposed Theory of Planned Behavior (DTPB) Model key variables of Attitude Belief, Subjective Norm (social influence) and Perceived Behavioral Control are decomposed them into specific belief dimensions of Perceived Ease Of Use and Perceived Usefulness.

7. *The Innovation Diffusion Theory*:
   The Innovation Diffusion Theory (IDT) is used to describe the innovation-decision process, which is largely fueled by the adoption of the IT system by the end user. The primary intention of this model is to provide an account of the manner in which any technological innovation moves from the stage of invention to its widespread deployment and use.

8. *The Expectation-Disconfirmation Theory*:
   The Expectation-Disconfirmation Theory (EDT) model focuses in particular on how and why user reactions change over time, when it comes to the final acceptance of an IT system.

9. *The Social Cognitive Theory*:
   The Social Cognitive Theory (SCT) model is based on the notion that environmental influences such as social pressures or unique situational characteristics, cognitive and other personal factors including personality as well as demographic characteristics key factors in the total adoption of an IT system.

10. *The Motivational Model*:

    The Motivational Model (MM) examines the key variable of extrinsic motivation and intrinsic motivation as it relates to the adoption of an IT system.

2.8.2 The Technology Acceptance Models

There are actually three variants of TAMs, which are described as follows:

1. *The Technology Acceptance Model (TAM1)*:

    This is the first version of TAM, and in fact, it is the most widely used today in terms of business adoption and academic research. This first variant made full usage of just pure psychological variables that affect Technology Acceptance. This model specifically states that the variables of Perceived Usefulness and Perceived Ease of Use ultimately determine the adoption and usage of an IT system, with the Intention To Use (Attitude) serving as the Mediator Variable of actual system use. This first variant is illustrated in Figure 2.1:

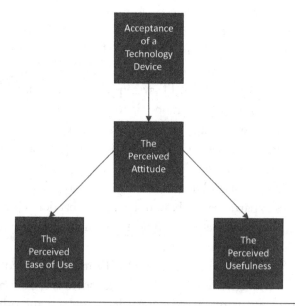

Figure 2.1   A graphic of the TAM, first version.

2. *The Unified Theory of Acceptance and Use of Technology*:
   This particular model is also known as the "UTAUT", which is deemed to be the third version of the original TAM model. This greatly enhanced model consists of eight new variables, which are as follows:
   - Performance Expectancy
   - Effort Expectancy
   - Social Influence
   - Facilitating Conditions
   - Gender
   - Age
   - Voluntariness
   - Experience

   This model can be seen in Figure 2.2:

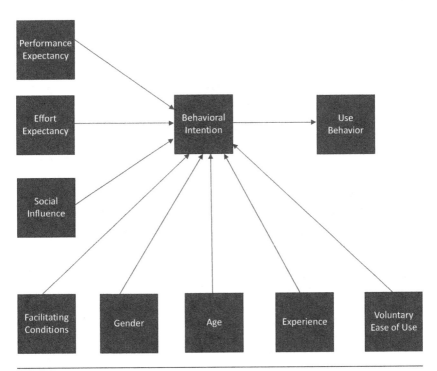

Figure 2.2   A graphic of the UTAUT model.

3. *The Technology Acceptance Model (TAM2)*:
   This is deemed to be the second model of TAM and is a theoretical extension of the original TAM model to address the following:
   - The impacts of Social Influence and Cognitive Instrumental processes;
   - How the effects of these determinants change with increasing user experience over time with the fully adopted IT system.

2.8.3 *The Survey Instrument*

Based on all of these theories, a research instrument was created, in which the final acceptance of a technological device is dependent upon the following two key variables:

- The Perceived Ease of Use of the Technological Device.
- The Perceived Usefulness of the Technological Device.

This "final" version of TAM is illustrated in Figure 2.3:

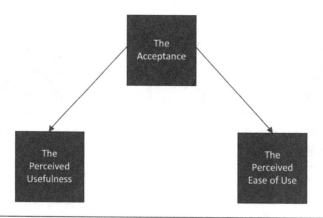

Figure 2.3 The graphic of the final TAM.

2.9 The Perceived Usefulness

The survey instrument questions that relate to The Perceived Usefulness are as follows:

1. My job would be difficult to perform without [XXXX].

2. Using [XXXX] will give me a greater control over my workflows.
3. Using [XXXX] will improve my overall work performance.
4. The [XXXX] addresses my job needs.
5. Using [XXXX] saves me time.
6. Using [XXXX] lets me to finish my job tasks quicker.
7. Using [XXXX] allows me to support the critical aspects of my job.
8. Using [XXXX] allows me to accomplish more work than I could ever before.
9. Using [XXXX] reduces the total amount of time that I spend on unproductive activities.
10. Using [XXXX] enhances the effectiveness of the productivity of my job.
11. Using [XXXX] enhances the quality of the work that I do.
12. Using [XXXX] increases my overall productivity.
13. Using [XXXX] makes it easier to do my job-related tasks.
14. Overall, I find that using [XXXX] is overall useful for my job.

In this case, any technological device can be used and is thus represented as "[XXXX]".

2.10 The Perceived Ease of Use

The survey instrument questions that relate to The Perceived Ease of Use are as follows:

1. I often become confused when [XXXX] for my job.
2. I tend to make more errors than normal when using [XXXX] for my job.
3. I need to often consult with the end user documentation and/or help guides when using [XXXX].
4. Working with the [XXXX] takes much more mental effort than is necessary.
5. Interacting with [XXXX] can be frustrating for me.
6. It is quite easy for me to recover from any errors that I make with [XXXX].
7. The [XXXX] is very rigid and inflexible to use.

8. It is easy to get the [XXXX] to what I want and/or need for it to do.
9. The [XXXX] works in ways that I have never expected before.
10. It is very cumbersome to use the [XXXX].
11. The interaction with the [XXXX] is quite easy to understand.
12. The [XXXX] makes it easy for me to remember tasks.
13. The [XXXX] does provide useful guidance when performing my job-related tasks.
14. Overall, I find that using the [XXXX] is quite easy to use.

In this case, any technological device can be used and is thus represented as "[XXXX]".

2.11 The Attitude

At this point, it is important to note that since The Attitude variable is composed of both The Perceived Ease of Use and The Perceived Usefulness, it is gauged by a ranking system, which is as follows:

> Good to Bad, where "1" is good and "5" is bad.
> Wise to Foolish, where "1" is Wise and "5" is Foolish.
> Favorable to Unfavorable, where "1" is Favorable and "5" is Unfavorable
> Beneficial to Harmful, where "1" is Beneficial and "5" is Harmful.
> Positive to Negative, where "1" is Positive and "5" is Negative.

It is important to note here that the ranking scale used is based on a numerical range from 1 to 5, with the "2", "3", and "4" holding intermediate values.

2.11.1 *The Survey Instrument for the Anti-Cyberbullying Mobile App*

While the TAM as it has been reviewed so far in the subsection relates primarily to the use of a technological device in terms of a work-related setting, the survey instrument questions and the ranking scale can be modified to fit any circumstance where there is acceptance of anything that is technology related. Therefore, it can also be modified in this regard to gauge the acceptance level of an Anti-Cyberbullying Mobile App. The next subsection will now review this.

2.12 The Perceived Usefulness

The survey instrument questions that relate to The Perceived Usefulness are as follows:

1. Keeping track of my child's online activities/behavior would be difficult without the "Anti-Cyberbullying Mobile App".
2. Using the "Anti-Cyberbullying Mobile App" will give me a greater control over my child's online activities/behavior.
3. Using the "Anti-Cyberbullying Mobile App" will improve my overall confidence that my child is safe.
4. The "Anti-Cyberbullying Mobile App" addresses the concerns that I have about my child's online activities/behavior.
5. Using the "Anti-Cyberbullying Mobile App" saves me time in keeping track of my child's online activities/behavior.
6. Using the "Anti-Cyberbullying Mobile App" lets me to address the concerns that I have about my child's online activities/behavior quicker.
7. Using the "Anti-Cyberbullying Mobile App" allows me to support my child's online activities/behavior.
8. Using the "Anti-Cyberbullying Mobile App" allows me to accomplish more tasks as they related to my child's online activities/behavior than I could ever before.
9. Using the "Anti-Cyberbullying Mobile App" reduces the total amount of time that I spend on unproductive activities related to my child's online activities/behavior.
10. Using the "Anti-Cyberbullying Mobile App" enhances the effectiveness of the productivity of keeping track of my child's online activities/behavior.
11. Using the "Anti-Cyberbullying Mobile App" enhances the quality of attention that I can give to my child as they conduct their online activities/behavior.

12. Using the "Anti-Cyberbullying Mobile App" increases my overall productivity as it relates to monitoring my child's online activities/behavior.

13. Using the "Anti-Cyberbullying Mobile App" makes it easier to do my tasks as they relate to my child's online activities/behavior.

14. Overall, I find that using the "Anti-Cyberbullying Mobile App" is useful for keeping tabs on my child's online activities/behavior.

2.13 The Perceived Ease of Use

The survey instrument questions that relate to The Perceived Ease of Use are as follows:

1. I often become confused when using the "Anti-Cyberbullying Mobile App".

2. I tend to make more errors than normal when using the "Anti-Cyberbullying Mobile App".

3. I need to often consult with the end user documentation and/or help guides when using the "Anti-Cyberbullying Mobile App".

4. Working with the "Anti-Cyberbullying Mobile App" takes much more mental effort than is necessary.

5. Interacting with the "Anti-Cyberbullying Mobile App" can be frustrating for me.

6. It is quite easy for me to recover from any errors that I make with the "Anti-Cyberbullying Mobile App".

7. The "Anti-Cyberbullying Mobile App" is very rigid and inflexible to use.

8. It is easy to get the "Anti-Cyberbullying Mobile App" to what I want and/or need for it to do.

9. The "Anti-Cyberbullying Mobile App" works in ways that I have never expected before.

10. It is very cumbersome to use the "Anti-Cyberbullying Mobile App".

11. The interaction with the "Anti-Cyberbullying Mobile App" is quite easy to understand.

12. The "Anti-Cyberbullying Mobile App" makes it easy for me to remember tasks.
13. The "Anti-Cyberbullying Mobile App" does provide useful guidance when performing my job-related tasks.
14. Overall, I find that using the "Anti-Cyberbullying Mobile App" is quite easy to use.

2.14 The Attitude

At this point, it is important to note that since The Attitude variable is composed of both The Perceived Ease of Use and The Perceived Usefulness, it is gauged by a ranking system, which is as follows:

1. Good to Bad, where "1" is good and "5" is bad.
2. Wise to Foolish, where "1" is Wise and "5" is Foolish.
3. Favorable to Unfavorable, where "1" is Favorable and "5" is Unfavorable.
4. Beneficial to Harmful, where "1" is Beneficial and "5" is Harmful.
5. Positive to Negative, where "1" is Positive and "5" is Negative. It is important to note here that the ranking scale used is based on a numerical range from 1 to 5, with the "2", "3", and "4" holding intermediate values.

2.15 The Scientific Limitations Presented by the Technology Acceptance Model

Although the Technology Acceptance Model has been in use for quite a long time (since 1985 to be exact), there have been a number of key criticism drawn to it. Probably the biggest one has been that TAM has only been used in a controlled environment, such as that of an academic one, where primarily students are surveyed. Thus, as other researchers have pointed out, it is quite difficult to extrapolate that into a real-world setting.

The second major criticism has been that the Technology Acceptance Model only considers the adoption of an IT system when the choice is made by free will, and not when it is mandatory. There have not been many studies with regard to the latter, and where they have been done, the results were negligible, in which Perceived Usefulness had more of an impact.

2.15.1 What It Means to Generative AI

These same criticisms will also hold true for an Anti-Cyberbullying Mobile App. But what will be different about the TAM in this regard is that it will have to take into considerations the impacts of Generative AI, as there will most likely be take into consideration as well when using the TAM as it is used to gauge The Acceptance level of the mobile app. The primary reason for this is that as Generative AI continues as rapidly as it is, there will be new functionalities that can be potentially embedded into the mobile app, such as geo-location (to keep better track of the child's whereabouts), facial recognition (to positively identify the perpetrator whom is launching the Cyberbullying attacks), content filtering mechanisms, providing more sophisticated alerts to the parents and school educators on a real-time basis, and having better control features.

Chapter 2 Resource

1. https://www.enablersofchange.com.au/what-is-the-technology-acceptance-model/

3

THE BASICS OF
GENERATIVE AI

As it has been reviewed throughout this book thus far, and in our previously published books on artificial intelligence (AI), Generative AI is now the "new norm" in the world of technology, and for that matter, even in just about every industry and market application that you can think of. But as it has also been stated, artificial intelligence (also known as "AI") can never replace the human brain. At best, all it can be used is for augmentation and automation, and to pick up on subtle clues that may not be so obvious to the human eye at first glance. This even applies to the world of Cyberbullying.

For example, Generative AI has two sides: positive and negative. In terms of the former, it can be used to help filter rogue contents in any kind or type of conversation that is used in Cyberbullying. From here, warnings can also be created and sent to the parents and school educators if the child is being Cyberbullied. But in terms of the latter, it can also be used to create and launch Cyberbullying attacks, as it also has been reviewed in the previous chapters.

Thus at this point, it is very important to review the basics of Generative AI. In our next chapter, we will do a deeper dive into some more advanced topics that are related to it.

3.1 How Generative AI Fits In

It is very important to note at this point that Generative AI is not by itself its own field. Rather, it is a subset of others that are associated with artificial intelligence; therefore, a review of them is thus pertinent. First, the question often arises: "Just what exactly is Artificial Intelligence?" A technical definition of it is as follows:

> Artificial intelligence represents a branch of computer science that aims
> to create machines capable of performing tasks that typically require

DOI: 10.1201/9781032666020-3

human intelligence. These tasks include learning from experience (machine learning), understanding natural language, recognizing patterns, solving problems, and making decisions.

(SOURCE: https://meng.uic.edu/news-stories/ai-artificial-intelligence-what-is-the-defini tion-of-ai-and-how-does-ai-work/)

In other words, the primary objective of artificial intelligence is to mimic as much as possible the thought and reasoning powers of the human brain, and apply that to a computer. This is a field that has been around for a long time, in fact going back to the early 1950s. Although the algorithms that were created and used back then are extremely primitive to what is being developed now, they are still very important, as they serve as the "backbone" for these very algorithms that are being developed now, and well into the future.

A subset of artificial intelligence that has evolved over time is what is known as "machine learning". It can be technically defined as follows:

Machine learning (ML) is the subset of artificial intelligence (AI) that focuses on building systems that learn – or improve performance – based on the data they consume.

(SOURCE: https://www.oracle.com/artificial-intelligence/machine-learning/what-is-machine-learning/)

As one can see from the definition, just like artificial intelligence, machine learning also makes use of datasets in order to compute a certain output, based on the query that has been presented to it. But the main distinction to be made between the two is that with the former, it is merely "Garbage In – Garbage Out", meaning it can only compute an output strictly based on the datasets that have been fed into it, and nothing more. But with the latter, the goal is to attempt to learn from the datasets that it has been trained upon, and try to compute an output that offers more than what has been fed into it. In other words, one of the other primary goals of machine learning is to have it "learn on its own", without much human intervention.

A perfect example of this is ChatGPT. Not only can it compute an output from the datasets that it has been trained on, but it can also search for other sources in order to further enhance it.

The next important subfield of artificial intelligence is what is known as "deep learning". A technical definition of it is as follows:

Deep Learning is a subfield of machine learning concerned with algorithms inspired by the structure and function of the brain called artificial neural networks.

(SOURCE: https://machinelearningmastery.com/what-is-deep-learning/)

To put it in simpler terms, and as it is stated in the definition up above, deep learning makes use of what is known as "neural networks". This is a much more sophisticated version of artificial intelligence, as "neurons" are used. As it relates to the human brain, the "neuron" is actually the cell of the human brain, and it forms the processes that are invoked when reasoning and thinking are involved. In fact, it has been estimated by many scientists that there are well over 86 billion neurons exist in just one human brain.

So to create a deep learning model, not only are the neurons are used, but they also reside at multiple layers which gives it ultra-sophistication. This is illustrated in Figure 3.1:

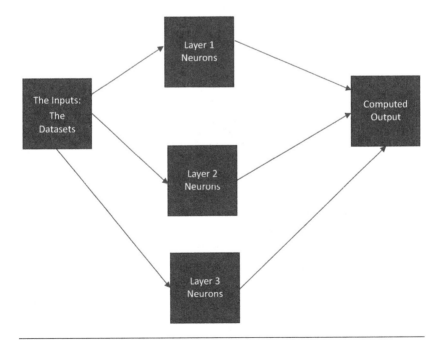

Figure 3.1   A graphic of a neural network.

Finally, we now come to what is known as "Generative AI". It is actually a subset of both machine learning and deep learning, and it too can be technically defined as follows:

> Generative AI enables users to quickly generate new content based on a variety of inputs. Inputs and outputs to these models can include text, images, sounds, animation, 3D models, or other types of data.

(SOURCE: https://www.nvidia.com/en-us/glossary/generative-ai/)

The distinction to be made between Generative AI and the other sub-fields of artificial intelligence just examined is that, and according to the definition, not only can different kinds of outputs be created, but the models that are also created from within the realms of Generative AI can even look at and examine other resources to compute the output, rather than simply relying upon the Datasets that it has been trained upon.

Once again, the prime example here is that of ChatGPT. When you submit a specific query to it, not only will it look up other resources that can be found on the internet, but it can also create the outputs in various formats, which include not only the text version of it, but also in terms of a video, an image, or even a sound file (this will be examined in more detail in the next section).

So as one can see, Generative AI is the culmination of not only just artificial intelligence but also machine learning and deep learning. This is illustrated in Figure 3.2:

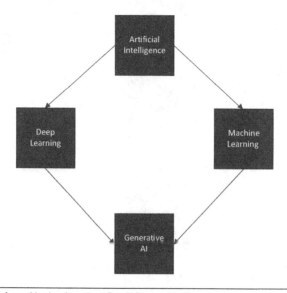

Figure 3.2   A graphic showing where Gen AI fits in with AI.

3.2 The Outputs That Are Created from Generative AI

The outputs that can be created from Generative AI were reviewed at a high level in the last section. In this section, we do a deeper dive into them. The outputs are as follows:

1. *The Text Generation*:

   In this instance, the output that is created in a tool like ChatGPT is merely a content-based one. The specific algorithms that are used to formulate the outputs are known as "The Transformer Architecture". This forms the constructs for the GPT3 and GPT4 algorithms. It can be technically defined as follows:

   The GPT algorithm, short for Generative Pre-trained Transformer algorithm . . . [it] is designed to generate coherent and contextually relevant text based on input prompts.

   (SOURCE: https://itexus.com/glossary/gpt-algorithm/#gref)

   It should be noted here that a prime catalyst for the GPT3 and GPT4 algorithms are the prompts that the end user creates when trying to get an answer (also the output) to their specific query. For instance, when you make use of Google as a search engine, you simply enter in some keywords or even a simple question. In response, and for that matter in just a matter of seconds, pages of resources come up that you can peruse through. But, these are not a specific answer to your query; rather, Google is letting you decide what is best. But with the GPT3 and GPT4 algorithms, the end user is actually getting a specific answer to their particular query. But given the ultra-sophistication of them, the end user has to construct a query that will yield the specific answer (also the output) that they are looking for. In fact, this has given rise to an entirely new field, which is known as "Prompt Engineering". This too can be technically defined as follows:

   Prompt engineering is the practice of designing and refining prompts – questions or instructions – to elicit specific responses from AI models.

   (SOURCE: https://www.datacamp.com/blog/what-is-prompt-engineering-the-future-of-ai- communication)

In other words, when using a tool such as ChatGPT, the more refined your query is, the better the result will be from your generated outputs. But it is important to keep in mind here that "Prompt Engineering" is not a subject that can be learned from an online course; rather, it is learned strictly through rote practice and repetition. The typical application for this is in content generation, such as creating a blog or an article of sorts. It can even be used to further check for the integrity of the grammar and syntax of the sentences for content that has already been created.

2. *The Image Generation*:

For this kind of output to be specifically created, Deep Learning Models are heavily utilized from within the realms of Generative AI. What makes something like ChatGPT different from artificial intelligence is that actual images can be used as a dataset to feed into the model. This can then be used to create an image to serve as the output. This kind of dataset can also be viewed as a "Qualitative"-based one, because it is not just a dataset based on pure numerical values. To create the images as outputs, ChatGPT also makes use of what are known as "General Adversarial Networks", or also known as "GANs" in short. In this case, a "Generator" is used to create the actual images based on what it has been trained on, and then the "Discriminator" is then used to determine if the created image is actually a fake or not. This is actually an iterative process that keeps repeating itself until an image is ultimately created that matches the need of the specific query. The applications for this are very numerous, as long as there is an image that needs to be created.

3. *The Audio Generation*:

In this kind of scenario, rather than creating a text or an image for the needed output, an audio file is created as the output. But what makes this different is that if the end user submits a specific query to ChatGPT, the output does not have to be an actual text. Rather, if the end user specifies that the output should be in the form of an audio file, then that is what ChatGPT will produce. In fact, the methodology that is used here is very similar to what is used in the

Virtual Personal Assistants of both Siri and Cortana. For instance, when you submit a query to either one of these, the generated output will always be that of an audio file. Some of the specific models that are used here by ChatGPT include "WaveGAN" and "Tacotron 2".

4. *The Video Generation*:

This kind of output that is generated and created by ChatGPT is actually deemed to be the most complex in terms of the processes that are involved with it. One of the primary reasons for this is that in this particular instance, there have to be many videos that have to be fed into whatever Generative AI Model that is used to compute the video-based outputs. They also must be of the correct nature; if not, the Generative AI Model will have to "guess" what the missing parts of the output will need to be. Although this entire process can transpire within minutes if not seconds, it can by its very nature take a lot of both computing and processing resources in order for the right kind of outputs to be created. But despite its complexity, video-based outputs are also deemed to be the most sophisticated, because it can make use of both computer vision and facial recognition, and the outputs that are yielded from this can serve a wide variety of market applications ranging from Forensics to Law Enforcement.

3.3 The Use of Natural Language Processing

Natural language processing, also known as "NLP" in short, was actually covered in our last book, entitled *Generative AI: Phishing and Cybersecurity Metrics*. But since NLP can also play a huge role in the development of tools to help combat Cyberbullying, it will be important to review here in this chapter as well. NLP can be technically defined as follows:

> Natural language processing (NLP) is a machine learning technology that gives computers the ability to interpret, manipulate, and comprehend human language.

(SOURCE: https://aws.amazon.com/what-is/nlp/)

In other words, NLP literally acts as the bridge between the human voice and the computer that is attempting to process it. This is illustrated in Figure 3.3:

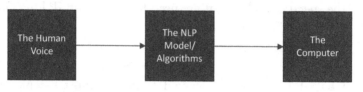

Figure 3.3   A graphic of the NLP.

It is important to note that there are a number of key concepts that are associated with NLP, which are reviewed in the next subsection:

3.3.1 The Concept of the N-Gram

1. *Tokenization*:
    This is the process where a sentence is broken into its individual words. These single words then become technically what are known as "Tokens". An example of this is as follows:
    "I love Artificial Intelligence"
    The above is the actual human voice.
    ["I"] ["love"] ["Artificial"] ["Intelligence"]
    The above is the "Tokenization", and each of the individual thus becomes a single "Token".
    "Tokenization" is primarily used so that large and complex human sentences can be broken down so that the NLP model and the corresponding algorithms can process them efficiently and effectively.

In the world of "Tokenization", another concept called the "N-grams" is also used. They are simply a mathematical sequence of the total number of "N Tokens" in any given sentence. They can be anything such as the actual words, spaces, punctuation marks, any type or kind of alphanumeric characters, phonemes, etc. There are different kinds of "N-Grams", which are also reviewed as follows:

1. *The Unigram*:
    This is also known technically as the "1-Gram", as its name implies. These are merely mathematical representations of each single word in a given sentence. So in this example:

"I love Artificial Intelligence"

["I"] ["love"] ["Artificial"] ["Intelligence"]

Each word in quotation marks thus becomes one, singular "Unigram".

2. *The Bigrams*:

This is also known technically as the "2-Gram", as its name implies. These are merely mathematical representations of two words "in a pair" in a given sentence. So in this example:

"I love Artificial Intelligence"

["I" "love"] ["Artificial" "Intelligence"]

Each two pair word in quotation marks thus becomes one, dual level "Bigram".

3. *The Trigram*:

This is also known technically as the "3-Gram", as its name implies. These are merely mathematical representations of each three word combination in a given sentence. So in this example:

["I" "love"] ["Artificial" "Intelligence"] ["very" "much"]

Every three word combination in quotation marks thus becomes one, trilevel "Trigram".

4. *The N-Gram in Language Modeling*:

This is a specialized algorithm that is used to compute the statistical probability of what a particular word actually means in the entire context of a sentence.

5. *The N-Gram in Text-Based Classification*:

This is a specialized algorithm in order to discern if a word has a positive or negative connotation to it. For example, in the examples up above, the word "love" represents a positive connotation, but if the word "despise" were to be substituted for it, it would thus have a negative connotation to it. This kind of algorithm is heavily used in a subfield of Generative AI which is called "Sentiment Analysis". This too can be technically defined as follows:

Sentiment analysis is the process of analyzing digital text to determine if the emotional tone of the message is positive, negative, or neutral.

(SOURCE: https://aws.amazon.com/
what-is/sentiment-analysis/)

6. *The N-Gram and Its Limitations*:

   In this situation, it is quite possible that an "N-Gram" may lose its ability to determine the context of a word, if the sentence is very long and convoluted. For example, in this sentence:

   "I love Artificial Intelligence, but I hate learning how to code".

   The algorithm will quickly pick up on the connotation of "love", but the chances for the word "hate" becomes lesser. Also, it is important to note here that if the algorithm is trying to decipher the context of individual words in a foreign language, it will not be able to do so unless the Generative AI Model has been specifically fed datasets of that particular language into it.

7. *The N-Gram in Out of Vocabulary Words*:

   The acronym for this kind of algorithm is known as the "OOV". In this case, the algorithm may not actually recognize a particular word in a given sentence, and thus it will assign what is known as a "Random Token" that is unique only to that particular word.

8. *The N-Gram and Smoothing*:

   This kind of algorithm will actually try to ascertain both the meaning and the contexts of words in a given sentence if there are too many of them in which connotations cannot be understood by the Generative AI Model.

3.3.2 *The N-Gram and Probabilistic Models*

As one can infer from the last subsection, NLP makes heavy use of theoretical statistics (for that matter, all of artificial intelligence makes use of this as well). But as it relates to the development and the usage of "N-Grams", there are a number of specialized models that have been created in this regard, which are as follows:

1. *The N-Gram Statistical Representation*:

   This is a kind of model where the individual words in a given sentence are further divided into an "N Sequence". For instance, a Unigram would be represented as "N=1", a

Bigram would be represented as "N=2", a Trigram will be represented as "N=3", and so forth.

2. *The Frequency Counting*:
   This is a kind of specialized model where it will count the number of occurrences of a particular word in a given sentence. It will also keep track of the total number of these occurrences for all the words that this model has been used for, and be used subsequently for training purposes.

3. *The Calculating Probabilities*:
   This model is also referred to as "N-Word Count". As this implies, this is where the model is applied in order to figure out the statistical probability of what the next word will be in a given sentence. For example, in the following sentence:

   "I love Artificial Intelligence, but I hate learning how to code".

   The model will try to compute the numerical probability of how likely that the word "Artificial" will appear after the word "love".

4. *The Smoothing*:
   Although this concept was reviewed in the last subsection of this chapter, "Smoothing" is used to actually compute the statistical probability that the model used here will come across an "N-Gram" that was not computed previously. The primary objective of this is to make the text of the spoken language error-free as possible, so that it can be easily processed by the NLP model and its corresponding algorithms.

5. *The Language Generation*:
   This model is similar to that of the "Frequency Counting" Model, but rather than trying to compute the statistical probability of just word appearing right after the preceding word, this kind of model will try to predict the statistical probabilities of random words appearing at unknown locations in any given sentence.

6. *The Hidden Markov Model*:
   This is yet another highly sophisticated model that is used in statistics today. It is particularly important for NLP in order to create what is known as a "Markovian" sequencing

of specific words in any given sentence. The Hidden Markov Model, also known as the "HMM", can attempt to predict with reasonable accuracy what the next series of sentences could be in a human conversation, based on what the "HMM" has seen before in terms of formulated sentences.

3.4 The Neural Network

Just like the same as for natural language processing, the concept of "neural networks" was also reviewed in detail in our previously published book entitled: *Generative AI: Phishing and Cybersecurity Metrics*. But just like NLP, it too plays a very important role in Generative AI, and even where Cyberbullying is concerned as well. It was also reviewed in detail in this chapter. To recap, this subset of artificial intelligence is used to mimic the human brain as much as possible by constructing layers of "Neurons" in its formulated model.

There are numerous kinds and types of neural network–based models, which are as follows:

1. *The Recurrent Neural Network*:
   This is also known as the "RNN". This is a kind of model where it is specifically designed to compute as well as process multiple sentences at one single point in time. What is unique about this is that it can provide a summary of what the connotation of a previous sentence and store that also, so that it can be used for subsequent training purposes. Another advantage that RNNs possess is that they can process sentences that are short and simple, and those that are long and complex as well. There are no numerical values that are placed onto the RNN, meaning it can process variable length sentences without too many problems. But, a key issue with the RNNs is that they have what is known as the "Vanishing Gradient Problem". This kind of mathematical procedure is typically used to update the statistical weights that are assigned to each of the neuron layers in the RNN model. While it can predict and assign these statistical weights for a short period of time, it has to be continually optimized so that it can do it all the time. This of course means more

training, and more of a subsequent drain of both memory and computing resources.

2. *The Long Short Term Memory*:

This is a form of a neural network model that was actually developed way back in 1997. What makes this different from other types and kinds of neural network models that are used in Generative AI is that it makes use of what is known as a "Memory Cell". This is where the model that is created can actually "forget" or "remember" sentences that are spoken and/or written in order to fit the specific application for the Generative AI Model that is being developed. This particular "Memory Cell" is controlled and regulated by three "Gates", which are as follows:

• *The Input Gate*:

This particular "Gate" controls the flow of new information and data that is fed into the Generative AI Model that is being used. It will also, depending upon the permutations that have been programmed into it, decide what pieces of information and data that the model will actually store. This will then be used for subsequent training purposes and for even computing the outputs.

• *The Forget Gate*:

This particular "Gate" will actually monitor and control the information and data that is already stored into the Generative AI Model, and if necessary, even delete them out from the "Memory Cells".

• *The Output Gate*:

This particular "Gate" controls the flow of information and data that is currently stored into the "Memory Cells" and which is also used to compute the outputs from the query that has been submitted to the Generative AI Model.

3. *The Gated Recurrent Unit*:

This is also technically known as the "GRU" in short. This kind of Artificial Intelligence Model is very similar to the last one just reviewed, but the primary difference, and in fact its main advantage, is that it is much simpler in design and is designed to have far fewer parameters. Thus, this makes the "GRU" a prime choice to be used where NLP is concerned in Generative AI. The "GRU" also only possesses two Gates, which are as follows:

- *The Reset Gate*:

This is used to determine what information and data need to be deleted from the "Memory Cell".

- *The Update Gate*:

This is used to determine how much new information and data should actually be stored into the "Memory Cell", in order to further optimize any future training that is involved for the Generative AI Model.

4. *The Encoder–Decoder*:

This is a type of neural network model that consists primarily of two distinct components:

- *The Encoder*:

This part actually processes the sentences, as they are fed into the Generative AI Model. It processes each word at a time, in an effort to fully understand both the context and meaning of them. These are then converted into what are known as "Context Vectors".

- *The Decoder*:

In this part, the "Context Vectors" are then fed into this, so that the outputs can then be formulated to the query that has been submitted to the Generative AI Model. Just like "The Encoder", "The Decoder" goes through each "Context Vector" one at a time, in an effort to guarantee the accuracy of the outputs.

5. *The Sequence to Sequence*:

This is a type of neural network model that is also very similar to that of the last model that was reviewed. This also consists of "The Encoder" and "The Decoder", but the key differences are as follows:

- It is designed and constructed making use of recurrent neural networks.
- It can handle sentence structures that are deemed to be of variable length.
- Although "The Encoder" and "The Decoder" work in almost the same fashion, another difference here is that

with the latter there is more of a direct relationship with the former so that the outputs that are computed from the Generative AI Model will be as accurate as possible.

It is important to note that with these last two neural network–based models, they are primarily optimized for analyzing a series of sentences that are shorter in nature. In order for them to analyze much longer sentences (e.g., a long and complex query that is submitted to a Generative AI Model), an add-on needs to be incorporated, which is technically known as the "Attention Mechanism". This add-on also permits for "The Encoder" and "The Decoder" to focus on specific parts of the sentences that are related to the query that has been submitted. It can also compute the importance of these specific areas of any given sentence, by making use of what is technically known as the "Attention Score".

6. *The Transformer Architecture*:

This is one of the newer forms of an Artificial Intelligence Model, which first appeared in 2017. It has been primarily designed to handle more complex cases and applications of NLP. This too has the same components of the last two models ("The Encoder" and "The Decoder") and the "Attention Mechanism", but it also consists of a brand new, fourth component, which is called the "Feed Forward Layer". It is used to capture the granular complexities syntax of much longer sentences. Also, when compared to the other neural network-based models in this subsection, "The Transformer Architecture" consists of the following advantages:

- It can analyze multiple sentences at a time, which is technically known as "Parallel Processing".
- Any kind or type of statistical weight that has been assigned can also be seen on a visual representation.
- It can also take very long sentences of a variable nature and analyze them at the same time.

In this chapter, we have examined some of the important basics of Generative AI. In the next chapter, we examine some of the more advanced features of it.

Chapter 3 Resources

1. https://meng.uic.edu/news-stories/ai-artificial-intelligence-what-is-the-definition-of-ai-and-how-does-ai-work
2. https://www.oracle.com/artificial-intelligence/machine-learning/what-is-machine-learning
3. https://machinelearningmastery.com/what-is-deep-learning
4. https://www.nvidia.com/en-us/glossary/generative-ai
5. https://itexus.com/glossary/gpt-algorithm/#gref
6. https://www.datacamp.com/blog/what-is-prompt-engineering-the-future-of-ai-communication
7. https://aws.amazon.com/what-is/nlp
8. https://aws.amazon.com/what-is/sentiment-analysis

Advanced Topics Into Generative AI

In the last chapter of this book, we reviewed some of the key concepts and basics that reside from within Generative AI. In this chapter, we look at more advanced topics, as they could also be incorporated into technologies and tools that are designed to help combat Cyberbullying. These topics will be reviewed in the following sections.

4.1 The Variational Autoencoder

These are also technically known as "VAEs" in short. This is also a variation of the Generative AI Model, and this kind of setup primarily makes use of what are known as "Autoencoders" and "Probabilistic Modeling". These kinds of Generative AI Models are designed to not only capture what they have learned in the past, but also apply that knowledge in making future predictions and computing the various outputs to the queries that are submitted to the model. This kind of Generative AI mechanisms consists of the following components:

1. *The Encoder*:
    This takes the datasets that not only has the Generative AI Model has been trained on, but what has been fed into as of recently. Unlike the other types and kinds of "Encoders" that have been reviewed previously in this book, this one does not represent the data as a fixed length. Rather, they are represented as a "Probability Distribution", using the concepts of statistics. This kind of approach allows for the "Encoder" to introduce a level of uncertainty into the Generative AI Model as it adapts and learns from new datasets over time.

2. *The Latent Space*:
    This is a "space" that has been specifically allocated into the Generative AI Model for those datasets that have been

DOI: 10.1201/9781032666020-4

deemed to be of a "Secondary Nature", and will not be used in the initial round of training for the Model. But rather, they will be called upon into the future if more datasets are needed to further optimize the Generative AI Model.

3. *The Reparameterization "Trick"*:

In this kind of scenario, rather than having the Generative AI Model learn directly from the datasets, hypothetical ones are created instead in order for it to train on. It should be noted that these kinds of datasets are also technically referred to as "Synthetic Data", and have been covered in more detail in our previously published books on artificial intelligence. It should also be noted here that even Generative AI itself can be used to create various forms of "Synthetic Data", which can be further used in other types and kinds of market applications.

4. *The Decoder*:

Although the concept of the "Decoder" has been reviewed previously in this book, as it applies to this situation, this functionality of the Generative AI Model will take a sample from the "Secondary Dataset" as just described, and attempt to actually "map" it back to the original datasets. In simpler terms, this is an attempt to use these "Secondary Datasets" to fill any voids or gaps that exist in the primary datasets upon which the Generative AI Model trains upon.

5. *The Loss Function*:

The VAE is created to be trained upon those datasets that have a lower statistical probability of being used by the Generative AI Model. There are two key concepts that are associated with this, which are as follows:

- *The Reconstruction*:

This functionality actually measures how well the "Synthetic Datasets" actually correlate with the "Primary Datasets" that are stored in the Generative AI Model.

- *The Regularization*:

This functionality helps to ensure that the "Synthetic Datasets" that have been produced and utilized are actually void of any gaps or holes that could potentially exist. The primary reason for doing this is that since these kinds of datasets are "manufactured data" and really have no real world value attached to them, they must, as much as possible, look like the real thing.

6. *The Generation and the Interpolation*:
   Once the Generative AI Model has actually been deemed to be fully trained, the VAE can then actually create new datasets, which are actually still "Synthetic" in nature, and from there, send it off to the "Decoder" in order to confirm and validate that indeed these new types of datasets can actually be subsequently used by the Generative AI Model not just for the purposes of training, but also for computing the outputs.

It is important to note here that VAEs are typically used by Generative AI Models which serve an application that is much more computing and processing intensive. For example, some typical applications include those of creating images, videos, and even compressing large datasets (these are typically known as "Big Data"). But on the contrary, VAEs can also skew the results of the outputs to a certain degree, as there could potentially be a lot of use of the "Synthetic Datasets".

4.2 The General Adversarial Network

These are also technically known as "GANs". While this was introduced in the last chapter, it is still important to touch on it again, but at a more technical level in this section. The GAN is actually a machine learning model of sorts, in which new datasets can be created from which it has learned on previously before. However, the GAN is not deemed to be as sophisticated as the VAE, as it can produce newer kinds of datasets only from the sources that it has been fed information and data from, whereas the VAE, given the extra number of components that it possesses, can produce newer types of datasets from other sources of information and data that it has not trained previously upon.

The GAN consists of four primary components, which are as follows:

1. *The Generator*:
   Although limited in sophistication to the "Encoder" (reviewed in the last section), the Generator still actually produce

newer types and kinds of datasets which closely parallel the datasets that have been fed previously into the Generative AI Model. But it should be noted here that at the initial outset, the datasets that have been produced by the "Generator" may not closely match the datasets that the Generative AI Model has been fed, and trained upon.

2. *The Discriminator*:

In technical terms, this component of the GAN is also known as the "Binary Classifier", in the sense that it can take the "real" datasets that have been ingested into the Generative AI Model and the produced datasets and from there, try to ascertain what is actually "fake" among the datasets. It should be noted here that this is actually an iterative process which keeps on cycling through the GAN so that eventually the reproduced or "fake" datasets will actually look like the real datasets.

3. *The Training*:

This is the actual iterative process just described. But, once this process is actually deemed to have been completed, the permutations that have been fed into the "Generator" are further modified, based on the number of cycles that are needed for the "fake" datasets to closely mirror and correlate with the real datasets. While this can possibly be done on an automated basis, it is highly advised that human intervention is also required at this point.

4. *The Equilibrium*:

It should also be noted that as the iterative process comes to a point of completion, this is technically known as the "State of Equilibrium". This simply means that it is, from the standpoint of statistics, for the discriminator to distinguish what is the "fake" and real data, even on a granular level. Further, it is at this point, that how and where the permutations should be modified needs to be taken into serious consideration.

Some of the typical applications for using a specific type of GAN include combining and superimposing various and different images into one main, composite image, and further augmenting datasets, when and where as needed, if not enough real datasets can be procured. It should also be noted that here that GANs are also heavily used in computer vision. This is yet another subfield of artificial intelligence in which a model tries to replicate the vision process of the human being. Although this is still far from being a reality, computer vision is quite useful in breaking down the pixels of a particular image into its most granular level of detail for further analysis. Also, GANs are used into computer vision when it comes to CCTV technology. In these particular cases, the camera also makes use of what is known as "facial recognition" to help positively identify any individual of interest that has been recorded by the camera. An illustration of computer vision can be seen in Figure 4.1:

Figure 4.1   An illustration of computer vision.

SOURCE: https://www.shutterstock.com/image-photo/selfdriving-3d-car-concept-person-steps-2198425187.

4.3 The Diffusion Model

Another recent advancement that has been made in the Generative AI is that of the "Diffusion Model". It is sophisticated in the sense that it actually draws upon the concepts of quantum mechanics and computer science. For example, whatever level of "noise" that may actually exist in the datasets (whether they are "fake" or real) can actually be converted into newer forms of datasets, that can also be ingested into the Generative AI Model. On a macro level, this process is done by actually reverse engineering the point where "noise" is first introduced in the datasets.

There are several key processes that are actually involved with a Diffusion Model, which as follows:

1. *The Noise Schedule*:

    This is where a sequencing of "noise" is first discovered by the Diffusion Model, and is thus represented as an upward slope where the least amount of "noise" is at the beginning of the gradient, and the most amount of noise is at the top of the gradient. This is illustrated in Figure 4.2:

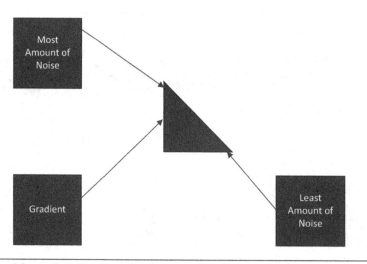

Figure 4.2   An illustration of an increasing amount of "noise" in the dataset.

There is also a distinct trade off here, in which the bottom of the gradient will represent datasets that are clear and fully optimized, and the top of the gradient represents those datasets that have almost no clarity to them whatsoever, and thus are not near to being fully optimized for the Generative AI Model to make use of and compute the outputs to queries that are presented to it.

2. *The Markov Chain*:

    This particular process makes use of the "Hidden Markov Model", as it was reviewed earlier in this book. But as it relates to the Diffusion Model, a "Markov Chain" represents a single point of space along the upward slope of the gradient. This space actually represents a point of "noise" and as the illustration up above illustrates, there will be more spaces like these, as you further along up the gradient. So in other words, there is a least amount of space at the very bottom of the gradient, but there will be most of spaces at the very top of the gradient. This is also illustrated in Figure 4.3:

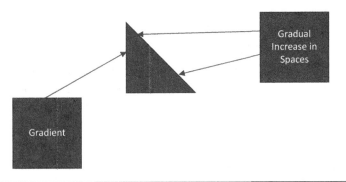

Figure 4.3 An illustration of mathematical spaces representing the "noise".

The ultimate goal of the "Markov Chain" is to "distort" all levels of "noise" as much as possible so that they will become unrecognizable to the Generative AI Model.

3. *The Conditional Modeling*:

   This process makes use of a specialized statistical technique that actually tries to make an "estimation" as to what the pieces of information and data could possibly look like at every space that becomes present along the gradient. It will also include in this analysis the information and data that were also present at previous steps. So in the end, when this technique reaches at the top of the gradient, this will be the total summation of all of the pieces of information and data which have "noise" embedded into them, and what their corresponding representations look like. The primary objective here is to see how much "noise" actually has to be distorted in the end.

4. *The Reverse Process*:

   This process actually occurs once the last step stops, or comes to a complete halt. The goal here is to ultimately remove all of the "noise" that have been detected and ascertained at all of the spaces along the gradient. Another objective here is to make these pieces of information and data resemble a realistic dataset as much as possible, so that it can be subsequently fed into the Generative AI Model for further usage.

One of the primary reasons why the Diffusion Model works so well in Generative AI is that they have been carefully designed so that any differences that reside between the "estimated" (or actually "hypothesized") data and any real datasets that have actually been observed are minimized as much as possible. It should also be noted here that typically the traditional Generative AI Models try to plot the "noise" to statistical distribution of the datasets, and the Diffusion Model totally avoids this from happening by minimizing the "noise" as much as possible. The end result of all of these is that any outputs that have been yielded by the Generative AI Model will be as authentic, realistic, and authentic as possible.

Also, Diffusion Models have been used where datasets need to be created that have a wide range of characteristics to them. For example, since a Generative AI Model can ingest both quantitative- and qualitative-based datasets, this diversity is greatly needed, and only a Diffusion Model can actually deliver this and make it actually happen.

4.4 The Types and Kinds of Diffusion Models

Now that we have reviewed what a Diffusion Model actually is, we can now advance one step further and provide an overview into the major categories of Diffusion Models. They are examined in this section.

1. *The Denoising Diffusion Probabilistic Model:*
   These are also referred to technically as "DPPM". This is mostly used where the datasets are those primarily of images. In this case, an image that is over pixilated and thus contains too much "noise" will eventually be eradicated through a defined process that resides from the "DDPM". This process is technically known as "maximum likelihood estimation", which tries to either get rid of the pixels in the image that are necessary, or simply close the distance between them so that the image will eventually become clearer, thus making it usable for the Generative AI Model.

2. *The Score-Based Diffusion Model*:

   This is also known as "SBM" in short. This kind of model makes use of what is known as a "Score Function". This is where the "SBDM" will actually estimate, or even measure the statistical probability that an image is actually a real one, and not a fake. In a way, the "SMB" is actually like an Artificial Intelligence Model of themselves, as they are trained by using an algorithm which is known as the "Adversarial Training". From this, the "SBM" can also be used to generate "fake" images that look like the real thing, although this is not its primary objective. The mathematical representation of the "SBM" is as follows:

   $$S(x) = x \ log \ P \ data(x)$$

3. *The Stochastic Differential Equation-Based Diffusion Model*:

   This is technically referred to as "SDE" in short. The reason why this particular model got the name that it did is that it uses an actual based "Stochastic Differential Equation" to actually generate images for the Generative AI Model, in a manner that is very much similar to all of the other processes that have been reviewed thus far in this chapter. The "SDE" also introduces a sense of "randomness" into the generated image, so that it will incorporate a sense of uniqueness into it. Further, this process is trained by using what is known as the "General Adversarial Training". It should also be noted that the reverse process can also happen with the "SDE", in that a generated is also broken down into various pieces of "noise". The mathematical representations of these are as follows:

   Creating an image from noise:

   $$Dx = f(x,t) \ * \ dt \ + \ g(t)^* \ dw$$

   Creating noise from an image:

   $$Dx = [f(x,t) - g^2(t)xLogPt(x)]^* \ dt = g(t)^* \ dw$$

4. *The Latent Representation Model*:

   This is also technically referred to as "LRM" in short. This is a specialized type of Diffusion Model that uses the architecture of a neural network (this was also reviewed earlier in this chapter, and this is where the concept of "Neurons" are used, and are thus represented by multiple layers in the actual Model). This approach is rather limited, as the neural network can only create what is known as a "Latent Image" strictly from the datasets that have been ingested into it and trained upon. These generated images are actually a collection of mathematical-based vectors, and this is stored into the Generative AI Model (assuming that the neural network component is also incorporated into it) so that future images can also be produced from this baseline. In this case, another type of neural network, called the "convolutional neural network" (also known as the "CNN"), can be used as well.

A "CNN" can be technically defined as follows:

Convolutional neural networks use three-dimensional data for image classification and object recognition tasks.

(SOURCE: https://www.ibm.com/topics/
convolutional-neural-networks)

A unique advantage of a "CNN" is that they can extract various features from an image, which could have a different scale length. Thus, this kind of model could also prove very useful facial recognition, which is a biometric modality that tries to confirm the identity of an individual by extracting the unique features from their face.

Also, along the with the "CNN", the "LRM" also makes use of a specialized statistical technique which is technically known as the "maximum likelihood estimation", also known as the "MLE" in short. One of the primary objectives of the "MLE" is to ascertain and implement the parameters for creating an image so that it can be subsequently used by the Generative AI Model. The "LRM" is mathematically represented as follows:

$$P_0 * \left(X_t - 1 \mid X_t \right)$$

5. *The Diffusion Process*:
   This is a process that also makes use of the "Markov Chain", but it is also deemed to be a "Probabilistic Process" from the standpoint of statistics. What is unique about this is that an

image is created from the raw datasets in a series of phases, which is directly visible to the end user that is trying to create them. Meaning, it can only move from one "Statistical State" to the next only step at a time, and not any quicker than that. To produce this, a concept called the "Diffusion Rate" is also applied, which simply means that throughout each "Statistical State", the generated image will not resemble anything like the datasets from which it has been derived from. Also, another technique called the "Gaussian Diffusion Process" can be used here.

6. *The Decoding Process*:
    This part of the Diffusion Model once again makes use of a neural network architecture, and in this case, an actual, real world image can be reverse engineered into the datasets from which it was originally derived from. A specialized statistical technique is also used here, which is known as the "mean squared error", or also known as "MSE" in short. Its primary objective in the Diffusion Model is to actually reduce the differences that have been found from the reverse engineered image and the actual, real world image.

4.5 The DALL-E 2

One of the cutting edge solutions that has occurred in Generative AI and especially to that of ChatGPT is having the functionality of taking a query that was submitted by, normal human language, and actually converting that directly to an image, as the output (if this has been created by the end user). The specific algorithm that drives is the "DALL-E 2", and it too was developed by OpenAI. How this process actually works is the focal point of this section, and it is as follows:

1. *The Inputs*:
    This is where the DALL-E 2 will take either an audio or a textual description of the image that is to be created. From there, it is then transferred into the Generative AI Model.

2. *The Encoding*:
    This is where the query that has been submitted (either via text or audio) is then further processed by the DALL-E

2 algorithm. At this point, it actually makes use of a specialized kind of neural network called the "Contrastive Language – Image Pre Training" also referred to technically as "CLIP". From here, the input that is provided by the end user becomes a mathematical-based, vector representation of it. The goal here at this step is to capture as much as possible the "semantic meaning" of the input.

3. *The Conversion*:

At this point in the process, the vector-based representations that have been produced by the "CLIP" are directed yet into another algorithm which is called the "Prior". This too is a Diffusion Model (or it can also be an Autoregressive one, based on the requirements that have been set forth onto the Generative AI Model), and this is deemed to be the first stage at where the submitted input actually starts getting converted over into an image of sorts. To do this, the "Prior" makes use of a Statistical-based, Probabilistic Model.

4. *The Generation*:

It is at this stage, after going through the required number of iterations in the last step, whatever has been produced by the "Prior" Algorithm are now thus transmitted over to the "Diffusion Decoder". This where it is used to convert all of the mathematical vectors that have been computed in the last step now become recognizable images, that can in the end be used as an output in order to satisfy the query that has been submitted to the Generative AI Model.

4.5.1 The Mechanics of the "CLIP"

For those who are technically oriented into the world of Generative AI, a common question that is asked is how the "CLIP" even trained at all, so that it can prove to be useful to a Generative AI Model? In reality, it consists of two major processes, which are known as the "Top Part" and the "Bottom Part".

4.6 The Top Part

In this particular phase, as its name implies, it is the top part of the image which is first processed by the "CLIP" (this is of course assuming that an actual image has been submitted as an input to the Generative AI Model). At this point, the following happens:

- The "CLIP" breaks down the image into certain "shared spots" and further examines for any relevant metadata as it relates to the image.
- The metadata then becomes statistically "joined" with one another, which allows for any information about the metadata to be shared easily.
- Ultimately, it is this statistically "joined" process that lets the Generative AI Model (such as that of ChatGPT) to fully understand the relationship between any text (whether written or spoken) and images that have been submitted to it.

4.7 The Bottom Part

In this particular phase, as its name implies, it is the bottom part of the image which is processed second by the "CLIP" (this is of course assuming that an actual image has been submitted as an input to the Generative AI Model). At this point, the following happens:

- The actual image generation happens in this phase, once again, using the appropriate Diffusion Model that has been selected to do this particular task.
- Any text inputs are also submitted to the "DALL-E 2" algorithm.
- The above-mentioned text inputs are further encoded by making use of the "CLIP Encoder", which thus creates high quality, mathematical representations of the text input. These are also technically known as the "CLIP Text Embeddings".
- These "Embeddings" are then processed through the "Prior" algorithm (which can once again be an Autoregressive-based or Diffusion-based Model). This algorithm then generates the "CLIP Image Embeddings", and these are used to mathematically correlate any visual context with that of the any textual-based context.
- At this last stage, the "CLIP Image Embeddings" are then further decoded by the appropriate "Diffusion Encoder", and it is here where the final images are thus created to satisfy the query that has been submitted to the Generative Model.

So, as one can see, it takes both processes to fully create an image, based on its upper and lower bounds. This process is illustrated in Figure 4.4:

Figure 4.4   An example of the DALL-E 2 algorithm.

SOURCE: https://www.shutterstock.com/image-photo/happy-puppy-welsh-corgi-14-weeks-2270841247.

4.8 The Stable Diffusion

This is also technically known as "Latent Diffusion". The one model that serves as the literal backbone for this is referred to as "Latent Diffusion Model", which is also known as "LDM" in short. It consists of the following components:

1. *The Diffusion Model in a Latent Space*:
    These kinds of sub-models are created and designed to work in a specific "latent space". Rather than applying any diffusion mechanics directly to the datasets that have been ingested into the Generative AI Model, they are applied into what is known as a "latent space". This then becomes the "Latent Representation of the Datasets".

2. *The Autoencoders/Latent Representatives*:
    As it has been reviewed throughout this chapter, the "Autoencoder" is a specialized kind of neural network, where it "Encodes" the ingested datasets and then compresses into a "Latent State". It is in this distinct region that any useful features of the datasets are captured and subsequently utilized.

3. *The Training and Optimization*:
   This is where the "LDM" is actually trained to transform and convert the ingested datasets into distinct "latent representations". A unique part about this process it can collect all kinds of "noises" that are present in the datasets, and later on, greatly reduce them so that they are unrecognizable to the Generative AI Model. But it is important to keep in mind that this part requires a lot of optimization, which may require a good amount of human intervention so that attention to detail can be observed and enforced.

4. *The Cross-Attention Layer*:
   This component is deemed to be a sophisticated "Add-On" to the LDM. This allows for it to understand all kinds of inputs that are put into the Generative AI Model, and thus, it can play a very important part in the creation of high quality and high resolution images as the output to the submitted query.

Finally, it should be added that there are a number of distinct advantages of using the LDM in a Generative AI Model, which are as follows:

1. *Processing Efficiency*:
   The LDM actually uses less computational power and processing power, when compared to other artificial intelligence-based models.

2. *A Balance Is Struck*:
   While the LDM is simple in design, it can also capture the most subtle and intricate details of an image, if it is used as part of a dataset that is ingested into the Generative AI Model.

3. *Diversity*:
   The LDM can take many different types and kinds of images (assuming also that they have been used as input) that are very diverse in nature, and in the end, create a robust image as an output that satisfies the nature query that has been submitted to the Generative AI Model.

Chapter 4 Resource

1. https://www.ibm.com/topics/convolutional-neural-networks

5

CONCLUSIONS

So far in this book, we have covered the following topics:

- An overview of Cyberbullying.
- The technological tools for combatting Cyberbullying.
- An overview of Generative AI.
- The advanced concepts of Generative AI.

In this last chapter of this book, we cover yet another important topic that relates to both Generative AI and Cyberbullying. That is, how to make the environment secure for the parents and school educators who want to use technology-based tools to help combat Cyberbullying against their kids and students. For example, a very key concern with Generative AI overall is how the data and information that it collects will be stored, and what are the risks of a data exfiltration from occurring?

There is no concrete answer to all of these yet, as Generative AI is still evolving, and new advances are being made to it on a very quick basis. But one area in which people who are involved in the design and creation of Generative AI Models is to ensure that the design of it is as robust and secure right from the beginning, and ensuring that the model is free from any known gaps or vulnerabilities before it is released into the production environment.

In fact, a similar area of concern as it relates to Cybersecurity is that of software development. Many software development teams still do not test their source code (which is used to create the actual, web-based application), thus making the delivered product to the client a very risky proposition to put into the production environment.

In an effort to allay this, many businesses today are using the concept of what is known as "DevSecOps". Simply put, this is a combination of the software development, IT security, and the operations, and they are all involved in the actual source code development. The idea here

DOI: 10.1201/9781032666020-5

is to have an "extra pair of eyes" to make sure that not only the source code is tested from a Cybersecurity standpoint, but also any gaps or vulnerabilities that have been discovered are quickly remediated.

This is especially true when the software development teams make use of what is known as "Application Programming Interface", or "API" in short. Essentially, these are repositories of source code which can be downloaded and further modified to fit the specific requirements of the web application project. By using APIs, software developers can bridge the backend (which is the database) and the front end (which is the Graphical User Interface, or "GUI") very quickly, without having to write even more source code.

While this is certainly advantageous in terms of cost savings and a timely delivery of the final product to the client, this also poses inherent Cybersecurity risks. For example, many of these APIs are available on what are known as "Open Source Libraries", and the vendors that maintain them do not keep these APIs updated with the latest patches and upgrades. As a result, the software developers fail to check for this also, because they assume that the APIs are safe to use to begin with. This has also been a catalyst for the inception of the "DevSecOps" team.

In a manner that is very similar to this, the concept of "LLMOps has thus been born". A technical definition of it is as follows:

> Large Language Model Ops (LLMOps) encompasses the practices, techniques and tools used for the operational management of large language models in production environments.

(SOURCE: https://www.databricks.com/glossary/llmops)

So in this case, the team that develops the Generative AI Model would also be coupled with the operations team, and very likely even with the IT security team to ensure that the model that has been created is not only robust, but that it has been thoroughly tested for any gaps or vulnerabilities (as well as remediated) before it is released into the production environment. So in this case, if a new Anti-Cyberbullying Mobile App was to be created that makes use of Generative AI, there is a chance that it will have been reviewed by the LLMOps team.

It is also important to note that the acronym of "LLM" stands for "Large Language Models". This concept was also reviewed in detail in our last book entitled: *Generative AI: Phishing and Cybersecurity Metrics*. But to review, it can be technically defined as follows:

> Large language models (LLMs) are a category of foundation models trained on immense amounts of data making them capable of understanding and generating natural language and other types of content to perform a wide range of tasks.

(SOURCE: https://www.ibm.com/topics/large-language-models)

When it comes to the creation of an Anti-Cyberbullying tool, LLMs will play a huge role, as it is the direct language of the perpetrator and the victim which will need to be carefully analyzed.

In fact, an overview of the LLMOps Framework is illustrated in Figure 5.1.

As one can see from the illustration earlier, using the concept of LLMOps in the development and deployment of Generative AI Models does have some distinct advantages which are as follows:

1. *Using Resources*:
   If the Generative AI Model is going to be deployed into a Cloud environment, such as that of Microsoft Azure, LLMOps can be used to make sure that only the necessary compute resources are used. Not only will this eliminate "sprawl", but it will also lead to cost savings down the road.

2. *Fine Tuning*:
   Generative AI Models will have to be optimized on a continual basis, and this is also where LLMOps can play a huge role that this is done correctly and safely from the very beginning.

3. *The Ethics*:
   As it has been mentioned throughout this book, while Generative AI can bring many benefits to the table, it can also be used for nefarious purposes. In terms of Cybersecurity, some of the greatest concerns are that of the creation of Deepfakes, and using a tool like ChatGPT can even be potentially used to create the source code in which to create malware. But there is also concern that Generative AI can also be used to create offensive content, especially when it comes to launching Cyberbullying attacks onto the social media platforms. The use of LLMOps can act as a secondary check here as well in order to mitigate the risk of these from happening.

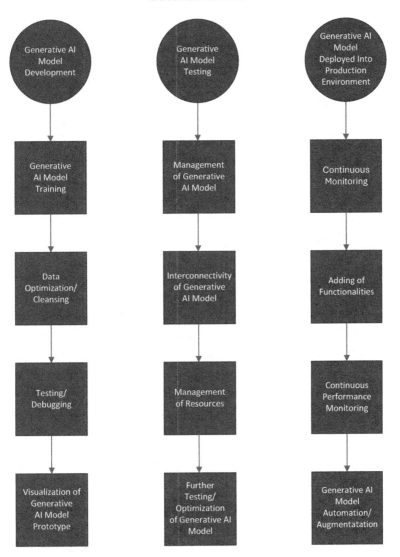

Figure 5.1 An illustration of the LLMOps processes.

4. *The Hallucinations*:

At its very fundamental core, Generative AI, despite all of its advancements, still follows the cardinal rule that is inherent in all of artificial intelligence: "Garbage In and Garbage Out". Because of this, Generative AI Models can also produce outputs that are greatly skewed. This is technically referred to as "Hallucinating" by the Generative AI Model. In these specific instances, the use of LLMOps can help to ensure that Generative AI Models that are created and deployed into the production environment are optimized and updated at all times in order to ensure that the outputs that have been created from it are as real and legitimate as possible.

5. *The Transparency*:

   In all of artificial intelligence, especially in that of Generative AI, there is a phenomenon known as "Black Box". This simply means that there is no transparency that is provided to the end user as to how the output was actually derived and what sources are used. There is now a greater movement in the AI community to have this kind of transparency, of course, without giving away the "Secret Sauce of the Algorithms".

6. *The Latency*:

   As Generative AI continues to evolve, there will be more demands that are placed on it as well in order to deliver the outputs in a very short time frame. In this regard, minutes will not do, the outputs will have to be derived within a matter of just a few seconds. Any delay in this regard is technically known as "latency", and to help alleviate this, the use of LLMOps can be integrated into the Generative AI Model development process to ensure that the algorithms are also optimized and trained with the latest datasets at all times.

7. *The Frameworks*:

   Unlike in the world of Cybersecurity, at the present time in the realm of artificial intelligence, there are no set of standards or best practices that businesses can follow in the creation and the development of the Generative AI Models. It is hoped that by making use of LLMOps, this will further catalyze the growth and adoption of these of standards and best practices for all to follow in the same fashion and manner.

The Appendix that follows this chapter will provide a further review of all of the Generative AI concepts that have been covered in this book.

Chapter 5 Resources

1. https://www.databricks.com/glossary/llmops
2. https://www.ibm.com/topics/large-language-models

Appendix A: Generative Deep Learning

GILLES PILON

Generative Deep Learning

There are two core concepts to understanding generative deep learning: generative modeling and deep learning.

Generative Modeling

"Generative modeling is a branch of machine learning that involves training a model to produce new data that is similar to a given dataset" (Foster 2023, p. 4).

Discriminative modeling: Imagine you are training a group of people to recognize the plays and poems written by William Shakespeare. You have a dataset of all of his works, as well as many works by many other authors. Each work in your dataset that was written by Shakespeare would be labeled with a 1 and all other works would be labeled with a 0. The group would learn that certain words, phrases, and styles are more likely to indicate that a work was written by Shakespeare. In supervised machine learning an algorithm learns from labeled data, called the training data, to make predictions or classifications. The group learns how to discriminate between these two groups (1, 0) and determines the probability that a new document shown to the group has a label of 1, that is, a work that was created by Shakespeare. This is called discriminative modeling.

Generative modeling: Imagine your training dataset contains only the works of William Shakespeare. Each work is called an observation. Each observation has many features, such as individual words or groups of letters. You train a generative model on this dataset to determine the rules of the complex relationships between the words in the works of Shakespeare. Now you can sample from this model to create new, realistic plays or poems that did not exist in the dataset, but look like they were created using the same rules as the original data. A generative model must also be probabilistic. We sample many different variations of the output; we do not get the same output every time, as we would with a discriminative model. A generative model has a random component that influences the outputs generated by the discriminative model. We want our model to generate new observations that look as if they were from the training dataset.

A discriminative model is easier to create, less demanding on the software and hardware, and has many use cases for solving problems. A model created to optimize the components of an industrial mixture is beneficial in a setting where the variability of the components changes constantly. A model to determine if a cancer is present in an image can speed up diagnosis. It is much easier to train a model to predict if a text file was written by a famous author or if an image was painted by a famous artist, than it is to build a model to generate text resembling that author or to create an image that resembles the artist. In the past decade software libraries have improved, new ones created, and hardware to solve specific computations in machine learning process data faster.

Generative modeling framework: This is a structure of our goals for our generative model (Foster 2023, pp. 10–11).

- *Training data*: Gather a dataset of observations that were generated according to an unknown distribution.
- *Modeling*: Build a generative model that mimics the unknown distribution and sample from the generative model to generate observations that appear to have come from the unknown distribution.

- *Accuracy*: Determine the accuracy of the generative model. A high accuracy means that a generative observation looks like it came from the unknown distribution.
- *Generation*: It should be easy to sample a new observation from the generative model.
- *Representation*: It should be possible to understand the high-level features in the dataset of observations as represented by the generative model. Each observation in the training data is described (mapped) using a lower-dimensional latent space.

Deep Learning

"Deep learning is a class of machine learning algorithms that uses multiple stacked layers of processing units to learn high-level representations from unstructured data" (Foster 2023, p. 23).

Structured data: These data are organized in a predefined format, like rows and columns in a spreadsheet. Supervised learning algorithms excel at finding patterns in these data because the features are well defined. Many machine learning algorithms require structured data.

Unstructured data: These data lack a predefined format, like text, images, or videos. These data may have a spatial structure (image), a temporal structure (audio, text), or both spatial and temporal structures (video). The individual observations (pixels, letters, frequencies, etc.) are very uninformative, that is, the data are granular. Various algorithms perform poorly with such granular data and the spatial or temporal structure. Deep learning is especially useful for unstructured data.

Deep learning is performed with any system that employs many layers to learn high-level representations of the training data. Most deep learning uses artificial neural networks that have multiple stacked hidden layers. Each layer contains units that are connected to the previous layer through a set of weights. The most common type of layer is the fully connected layer (also known as the dense layer) which means every unit in a layer is directly connected to every unit in the previous layer. There are several types of artificial neural networks. We will consider the multilayer perceptrons and the convolutional neural network.

Multilayer perceptrons (MLP): All adjacent layers are fully connected. The input is transformed by each layer in turn until it reaches the output layer. Each unit applies a transformation to its inputs and passes the output to the next layer. A single unit in the final layer outputs the probability that the original input belongs to a specific category. The transformation involves a weighted sum of the inputs. The deep neural network finds the set of weights for each layer by training the network. Text, audio, or images are processed through the network and the predicted outputs are compared to the truth. If there are errors in the prediction these are propagated backward through the network and the weights are adjusted to find those which improve the prediction; this is called backpropagation. An artificial neural network learns features from the training data without human guidance in order to minimize the prediction error.

Convolutional neural network (CNN): This type of neural network takes into account the spatial structure of the training data. This is particularly useful for images. The image is converted to a single vector before passing it to the dense layer. A convolution is a mathematical operation of sliding a filter across the input data (typically an image) and performing element-wise multiplication. The result is a new value that captures how well the filter matches the features in a specific region of the input. A convolutional layer takes the input and applies convolutions with multiple filters to generate feature maps. It is a collection of filters and the values stored in the filters are the weights learned through training. Each filter in the layer detects a specific feature in the input data. By stacking multiple convolutional layers, a CNN can learn increasingly complex features from the input image, ultimately leading to object recognition or other image analysis tasks.

Methods of Generative Modeling

Before delving into the methods of generative modeling, there is one key concept to understand. A *probability density function* is used to specify the probability of a random variable falling within a particular range of values, as opposed to taking on any one value. Instead of a single guess, the model provides a range of possibilities with their corresponding likelihoods.

There are two different approaches of generative modeling:

Implicit density models: These produce a random (also called stochastic) process of generating data. These models do not calculate the probability density function.

Explicit density models: These constrain the model-building
process so that the probability density function is easier to
calculate.

Generative Adversarial Networks

Generative adversarial network (GAN) is an implicit density model.
A GAN is a battle between two adversaries. Imagine you have two art
specialists, one a talented counterfeiter (generator) and the other an art
critic (discriminator). They are locked in an artistic battle.

The counterfeiter (generator) keeps creating new forgeries, trying
to mimic the style of famous artists (data). The critic (discriminator)
examines each forgery and tries to determine if it is a real painting or
a fake. Over time, the counterfeiter gets better at creating convincing fakes, while the critic becomes a sharper judge. This competition
pushes both to improve. The counterfeiter learns the subtleties of the
artist's style, and the critic hones their ability to detect even the most
minor discrepancies.

This is how a GAN works. One neural network (generator) creates new
data (like images or music) based on training data. The other network (discriminator) tries to distinguish the generated data from real data. Through
this adversarial process, the generator becomes adept at producing data
that closely resembles the real thing, while the discriminator keeps the
generator on its toes by constantly improving its detection abilities. This
makes GANs powerful tools for generating new, realistic data for various
applications, from creating new images to composing novel music pieces.
They constantly push the boundaries of what artificial intelligence can
create, blurring the lines between the artificial and the real.

Variational Autoencoder

Variational autoencoder (VAE) is an explicit density model. VAEs can
be thought of as artistic compressors. Imagine you have a box filled
with different kinds of toys (training data), and you want to compress

them into a smaller box (latent space) while still being able to recreate them (generate new toys) later.

A regular autoencoder would simply shrink the toys down and then try to inflate them back to their original size. This process is messy and the resulting toys might be blurry or miss some details.

A variational autoencoder is more sophisticated. It encodes the toys into a special kind of compressed space that captures not only their size but also some of their key features. It is like having a box filled with colored blocks (latent space) where each block represents a specific type of toy (a red block for cars, a blue block for dolls, a green block for plush toys). By sampling from this box of blocks and using a decoder, a VAE can generate new, never-before-seen toys that share characteristics with the originals. This makes them useful for tasks like creating realistic images of faces or coming up with new music samples that fit a particular style.

Energy-based Model

Energy-based model (EBM) is an explicit density model. EBMs are like happiness detectors for data. Imagine a landscape with hills and valleys, where the height of the ground represents the model's "energy" for a particular data point. Data points that fit the model well (real or desirable) would be in valleys with low energy. Data points that do not fit (outliers or noise) would be on hills with high energy.

The core idea of EBMs is to define a function that assigns an energy score to every possible data point. The model then learns to adjust this function so that low energy regions correspond to the kind of data it is trying to model. Think of it like sculpting a clay model. You keep smoothing the clay (adjusting the energy function) to create a landscape with deep valleys where you want the features of your model to reside. Data points that land in these valleys are considered likely or desirable by the model.

EBMs are a powerful tool for creating realistic images or music. By sampling data points from low energy regions, they can produce new data that adheres to the patterns learned from the training data. Additionally, their probabilistic nature allows them to estimate the likelihood of any generated data point.

Diffusion Models

This is an explicit density model. Imagine you have a clear image and slowly add random noise to it, making it increasingly blurry and unrecognizable over time. This noise addition process is essentially what a diffusion model does in reverse.

The model is first trained on real data by learning this noise addition process. It essentially learns how to corrupt clear images with controlled noise step-by-step. Then, to generate new data, the model starts with pure noise and reverses this corruption process one step at a time. Think of it like slowly removing fog from a landscape photo. With each step, the model removes some noise, revealing more and more details of the underlying image. By learning the process of adding noise, the model can learn how to remove it effectively, ultimately creating a new image that resembles the training data.

Diffusion models are particularly adept at capturing complex structures and realistic details in data like images or audio. Their ability to learn the data distribution through a denoising process makes them powerful tools for various generative tasks.

Autoregressive Models

This is an explicit density model. An autoregressive model is like a mathematical equation that takes a series of past values as inputs and outputs a prediction for the next value in the sequence. By analyzing historical data, the model learns these underlying patterns and uses them for forecasting. Autoregressive models are like fortune tellers who predict the future based on the past. Imagine a fortune teller who uses tarot cards (past data points) to predict your upcoming week (future outcomes). They analyze the cards you draw (past values) and based on the patterns they have seen before (relationships between past and future events), make predictions about your future (future values).

In statistics and machine learning, autoregressive models use past values of a time series to predict the next one. They are particularly useful for data that exhibits some dependence on its history, like stock prices or weather patterns. However, autoregressive models have limitations. Just like a fortune teller cannot predict everything, these models can struggle with significant changes or unexpected events. They assume the future will resemble the past to some extent, which might not always be true.

Normalizing Flow Models

This is an explicit density model. Imagine you have a simple distribution of data, like a bunch of balls clustered in the center of a sandbox (latent space). A normalizing flow model can transform this simple distribution into a more complex one, like sculpting the sandbox into intricate shapes (complex data). Think of it as a flexible mold. You can start with a simple shape and keep applying transformations to create more intricate and varied forms, all while maintaining the ability to go back to the original shape if needed. This makes normalizing flow models powerful for generating realistic and diverse data while providing valuable insights into the probability of the generated data.

The normalizing flow model achieves this by applying a series of reversible functions, like pushing and pulling the sand with tools. Each function modifies the distribution in a controlled way, gradually shaping it into the desired form. The key here is that these functions are invertible, so you can always reverse the steps to get back to the original distribution. This invertibility is crucial because it allows the model to not only generate new data points (like scooping out new shapes in the sandbox) but also efficiently calculate the probability of any data point it generates. This makes normalizing flows attractive for tasks where understanding the likelihood of generated data is important.

Deep Learning Software

Creating deep learning models requires powerful, feature-rich software frameworks. TensorFlow with Keras are exceptionally good Python packages.

TensorFlow

It is an open-source software package for machine learning and artificial intelligence. It can be used across a range of tasks but has a particular focus on training and inference of deep neural networks. See https://github.com/keras-team/keras.

Keras

It is an open-source software for deep learning with a high-level application programming interface (API) that makes it very simple to train and run artificial neural networks. It provides numerous methods that can be combined to create very complex deep learning applications. It runs on top of TensorFlow and other packages. See https://github.com/tensorflow/tensorflow.

Appendix A References

Foster, David (2023). *Creative Deep Learning: Teaching Machines to Paint, Write, Compose, and Play.* 2nd ed. Sebastopol, CA: O'Reilly Media, Inc., p. 456.
Google Brain (2024a). *Keras.* https://github.com/keras-team/keras
Google Brain (2024b). *TensorFlow.* https://github.com/tensorflow/tensorflow

Index

Printed in the United States
by Baker & Taylor Publisher Services